ENERGY SCIENCE, ENGINEERING AND TECHNOLOGY

ENERGY STORAGE: ISSUES AND APPLICATIONS

JONATHAN M. BOWEN
EDITOR

Nova Science Publishers, Inc.
New York

For permission to use material from this book please contact us:
Telephone 631-231-7269; Fax 631-231-8175
Web Site: http://www.novapublishers.com

NOTICE TO THE READER

LIBRARY OF CONGRESS CATALOGING-IN-PUBLICATION DATA
Energy storage : issues and applications / editor, Jonathan M. Bowen.
p. cm.
Includes bibliographical references and index.
ISBN 978-1-61209-517-2 (hardcover)
1. Energy storage. I. Bowen, Jonathan M.
TJ165.E5255 2011
621.042--dc22
2010051529

Published by Nova Science Publishers, Inc. † New York

ENERGY SCIENCE, ENGINEERING AND TECHNOLOGY

ENERGY STORAGE: ISSUES AND APPLICATIONS

ENERGY SCIENCE, ENGINEERING AND TECHNOLOGY

Additional books in this series can be found on Nova's website under the Series tab.

Additional E-books in this series can be found on Nova's website under the E-books tab.

CONTENTS

PREFACE

Both solar photovoltaics and wind energy have variable and uncertain outputs, which are unlike the dispatchable sources used for the majority of electricity generation in the United States. The variability of these sources has led to concerns regarding the reliability of an electric grid, as well as the cost of integrating large amounts of variable generation into the electric grid. This new book explores the role of energy storage in the electricity grid, focusing on the effects of large-scale deployment of variable renewable sources, primarily wind and solar energy.

Chapter 1- Renewable energy sources, such as wind and solar, have vast potential to reduce dependence on fossil fuels and greenhouse gas emissions in the electric sector. Climate change concerns, state initiatives including renewable portfolio standards, and consumer efforts are resulting in increased deployments of both technologies. Both solar photovoltaics (PV) and wind energy have variable and uncertain (sometimes referred to as "intermittent")[1] output, which are unlike the dispatchable sources used for the majority of electricity generation in the United States. The variability of these sources has led to concerns regarding the reliability of an electric grid that derives a large fraction of its energy from these sources as well as the cost of reliably integrating large amounts of variable generation into the electric grid. Because the wind doesn't always blow and the sun doesn't always shine at any given location, there has been an increased call for the deployment of energy storage as an essential component of future energy systems that use large amounts of variable renewable resources. However, this often-characterized "need" for energy storage to enable renewable integration is actually an economic question. The answer requires comparing the options to maintain the required system reliability, which include a number of technologies and changes in operational practices. The amount of storage or any other "enabling" technology used will depend on the costs and benefits of each technology relative to the other available options.

Chapter 2- As renewable electricity becomes a larger portion of the electricity generation mix, new strategies will be required to accommodate fluctuations in energy generation from these sources. One of the primary strategies proposed for integrating large amounts of renewable energy is using energy storage to absorb excess electricity-generating capacity during times of low demand and/or high rates of generation by renewable sources and then reconverting this stored energy into electricity during periods of high demand and/or low renewable generation.

Chapter 3- Recent and ongoing improvements in thermal solar generation technologies coupled with the need for more renewable sources of energy have increased interest in concentrating solar thermal power (CSP). Unlike solar photovoltaic (PV) generation, CSP uses the thermal energy of sunlight to generate electricity. Two common designs of CSP plants—parabolic troughs and power towers—concentrate sunlight onto a heat-transfer fluid (HTF), which is used to drive a steam turbine. An advantage of CSP over non-dispatchable renewables is that it can be built with thermal energy storage (TES), which can be used to shift generation to periods without solar resource and to provide backup energy during periods with reduced sunlight caused by cloud cover.[1] The storage medium is typically a molten salt, which has extremely high storage efficiencies in demonstration systems.

In: Energy Storage: Issues and Applications
Editors: Jonathan M. Bowen
ISBN: 978-1-61209-517-2

Chapter 1

The Role of Energy Storage with Renewable Electricity Generation*

Paul Denholm, Erik Ela, Brendan Kirby and Michael Milligan

Acknowledgments

The authors wish to thank the following people for their review and other contributions: Doug Arent, Easan Drury, Vahan Gevorgian, Michelle Kubik, Jim Leyshon, Sean Ong, and Brian Parsons of the National Renewable Energy Laboratory (NREL); Jonah Levine of the University of Colorado; Mark O'Malley of the Electricity Research Center, University College Dublin; and Samir Succar of the Natural Resources Defense Council.

List of Acronyms

AC	alternate current
AGC	automatic generation control
AS	ancillary services

* This is an edited, reformatted and augmented edition of a National Renewable Energy Laboratory publication, Report NREL/TP-6A2-47187, dated January 2010.

CAES	compressed-air energy storage
CCGT	combined-cycle gas turbine
CCR	capital charge rate
CSP	concentrating solar power
DC	direct current
ERCOT	Electric Reliability Council of Texas
EVs	electric vehicles
GW	gigawatt
ISO	independent system operators
LaaR	Load Acting as a Resource (program)
MW	megawatt
NERC	North American Electric Reliability Corporation
PHS	pumped hydro storage
PV	photovoltaics
RTO	regional transmission organization
RE	renewable energy
SMES	superconducting magnetic energy storage
T&D	transmission and distribution
V2G	vehicle to grid
VG	variable generation

1. INTRODUCTION

Renewable energy sources, such as wind and solar, have vast potential to reduce dependence on fossil fuels and greenhouse gas emissions in the electric sector. Climate change concerns, state initiatives including renewable portfolio standards, and consumer efforts are resulting in increased deployments of both technologies. Both solar photovoltaics (PV) and wind energy have variable and uncertain (sometimes referred to as "intermittent")[1] output, which are unlike the dispatchable sources used for the majority of electricity generation in the United States. The variability of these sources has led to concerns regarding the reliability of an electric grid that derives a large fraction of its energy from these sources as well as the cost of reliably integrating large amounts of variable generation into the electric grid. Because the wind doesn't always blow and the sun doesn't always shine at any given location, there has been an increased call for the deployment of energy storage as an essential component of future energy systems that use large amounts of variable renewable resources. However, this often-characterized "need" for energy storage to enable renewable integration is actually an economic question. The answer requires comparing the options to maintain the required system reliability, which include a number of technologies and changes in operational practices. The amount of storage or any other "enabling" technology used will depend on the costs and benefits of each technology relative to the other available options.

To determine the potential role of storage in the grid of the future, it is important to examine the technical and economic impacts of variable renewable energy sources. It is also important to examine the economics of a variety of potentially competing technologies including demand response, transmission, flexible generation, and improved operational practices. In addition, while there are clear benefits of using energy storage to enable greater penetration of wind and solar, it is important to consider the potential role of energy storage in relation to the needs of the electric power system as a whole.

In this chapter, we explore the role of energy storage in the electricity grid, focusing on the effects of large-scale deployment of variable renewable sources (primarily wind and solar energy). We begin by discussing the existing grid and the current role that energy storage has in meeting the constantly varying demand for electricity, as well as the need for operating reserves to achieve reliable service. The impact of variable renewables on the grid is then discussed, including how these energy sources will require a variety of enabling techniques and technologies to reach their full potential. Finally, we evaluate the potential role of several forms of enabling technologies, including energy storage.

2. Operation of the Electric Grid

The operation of electric power systems involves a complex process of forecasting the demand for electricity, and scheduling and operating a large number of power plants to meet that varying demand. The instantaneous supply of electricity must always meet the constantly changing demand, as indicated in Figure 2.1. It shows the electricity demand patterns for three weeks for the Electric Reliability Council of Texas (ERCOT) grid during 2005.[2] The seasonal and daily patterns are driven by factors such as the need for heating, cooling, lighting, etc. While the demand patterns in Figure 2.1 are for a specific region of the United States, many of the general trends shown in the demand patterns are common throughout the country. To meet this demand, utilities build and operate a variety of power plant types. Baseload plants are used to meet the large constant demand for electricity. In the United States, these are often nuclear and coal-fired plants, and utilities try to run these plants at full output as much as possible. While these plants (especially coal) can vary output, their high capital costs, and low variable costs (largely fuel), encourage continuous operation. Furthermore, technical constraints (especially in nuclear plants) restrict rapid change in output needed to follow load. Variation in load is typically met with load-following or "cycling" plants. These units are typically hydroelectric generators or plants fueled with natural gas or oil. These "load-following" units are further categorized as intermediate load plants, which are used to meet most of the day-to-day variable demand; and peaking units, which meet the peak demand and often run less than a few hundred hours per year.

In addition to meeting the predictable daily, weekly, and seasonal variation in demand, utilities must keep additional plants available to meet unforeseen increases in demand, losses of conventional plants and transmission lines, and other contingencies. This class of responsive reserves is often referred to as operating reserves and includes meeting frequency regulation (the ability to respond to small, random fluctuations around normal load), load-forecasting errors (the ability to respond to a greater or less than predicted change in demand), and contingencies (the ability to respond to a major contingency such as an unscheduled power plant or transmission line outage) (NERC 2008).[3] Both frequency regulation and contingency reserves are among a larger class of services often referred to as ancillary services, which require units that can rapidly change output. Figure 2.2 illustrates the need for rapidly responding frequency regulation (red) in addition to the longer term ramping requirements (blue). In this utility system, the morning load increases smoothly by about 400 megawatts (MW) in two hours. During this period, however, there are rapid short-term ramps of +/- 50 (MW) within a few minutes.

Because of the rapid response needed by both regulation and contingency reserves, a large fraction of these reserves are provided by plants that are online and "spinning" (as a result, operating reserves met by spinning units are sometimes referred to as spinning reserves.)[4] Spinning reserves are provided by a mix of partially loaded power plants or responsive loads. The need for reserves increases the costs and decreases the efficiency of an electric power system compared to a system that is perfectly predictable and does not experience unforeseen contingencies. These costs result from several factors. First, the need for fast-responding units results in uneconomic dispatch – because plants providing spinning reserve must be operated at part load, they potentially displace more economic units.[5] (Flexible load-following units are often either less efficient or burn more expensive fuel than "baseload" coal or nuclear units.) Second, partial loading can reduce the efficiency of individual power plants. Finally, the reserve requirements increase the number of plants that are online at any time, which increases the capital and O&M costs.

Figure 2.3 provides a simplified illustration of the change in dispatch (and possible cost impacts) needed to provide operating reserves. The figure on the left shows an "ideal" dispatch of a small electric power system. Two baseload units provide most of the energy, while an intermediate load and two peaking units provide load following. In the "ideal" dispatch, it is possible that the intermediate load unit cannot rapidly increase output to provide operating reserves. Furthermore, during the transition periods when the load-following units are nearing their full output – but before additional units are turned on – there may be insufficient capacity left in the load-following units to provide necessary operating capacity for regulation or contingencies. A dispatch that provides the necessary reserves is provided on the right. In this case, lower-cost units reduce output to accommodate the more flexible units providing reserves. This increases the overall cost of operating the entire system.

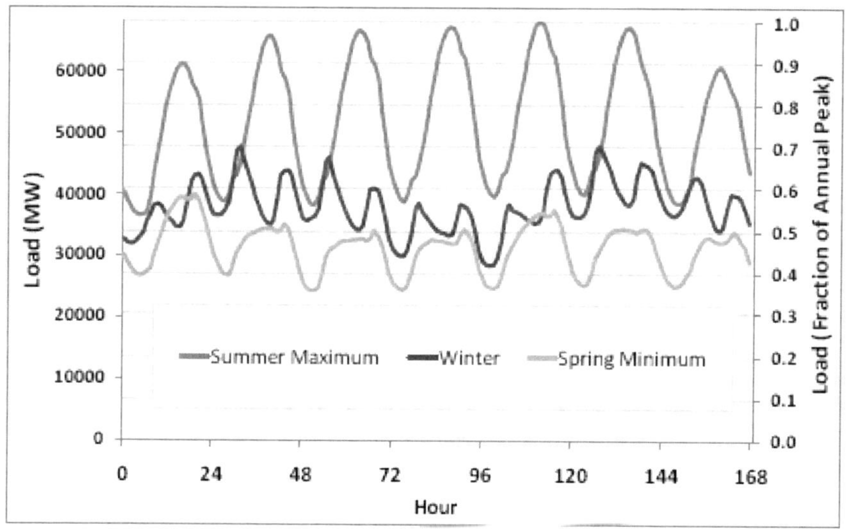

Figure 2.1. Hourly loads from ERCOT 2005

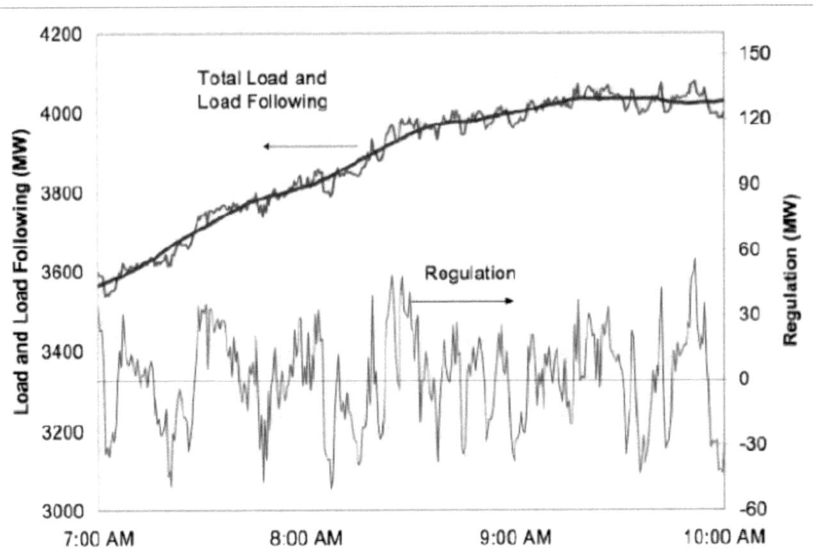

Figure 2.2. System load following and regulation. Regulation (red) is the fast fluctuating component of total load (green) while load following (blue) is the slower trend (Kirby 2004)

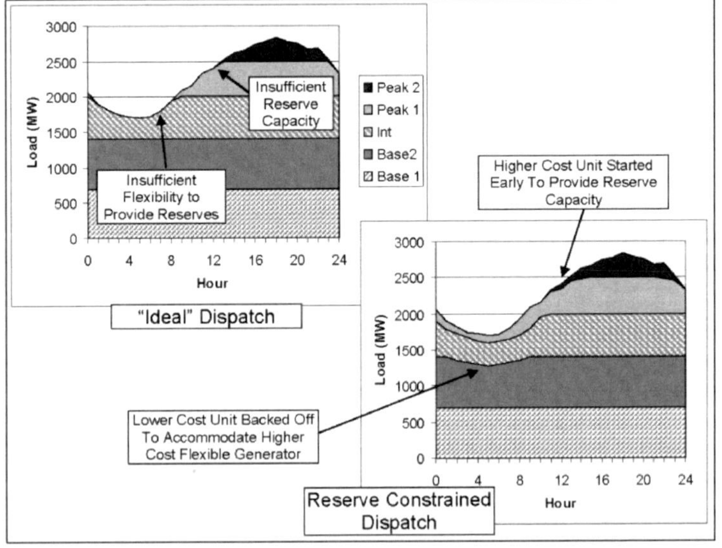

Figure 2.3. Optimal and reserve constrained dispatch

The need for operating reserves and the large variation in demand restricts the contribution from low-cost baseload units and increases the need for units that can vary output to provide both load-following and ancillary services. As a consequence, utility operators have long pursued energy storage as one potential method of better utilizing baseload plants and providing an alternative to lower efficiency thermal generators for meeting variations in demand.

3. Electricity Storage in the Existing Grid

The challenges associated with meeting the variation in demand while providing reliable services has motivated historical development of energy storage. While a number of pumped hydro storage (PHS) plants[6] were built in the United States before 1970, significant interest, research, and funding for new storage technologies began in the early 1 970s, associated with dramatic increases in oil prices. This period also saw the largest deployment of PHS based on its competitive economics compared to alternative sources of intermediate load and peaking energy.

3.1. Development of Energy Storage in Regulated Markets

Deployment of energy storage is dependent on the economic merits of storage technologies compared to the more conventional alternatives used to follow load. Before the advent of low-cost, efficient gas turbines now typically used to follow load and provide reserves, utilities often relied on oil- and gas-fired steam turbines (and hydroelectric dams where available). In the 1 970s, dramatic price increases in oil and natural gas occurred, along with concerns about security of supply. This led to the Powerplant and Industrial Fuel Use Act, restricting use of oil and gas in new power plants (EIA 2009b). Utilities expected to bring online many new coal and nuclear plants to meet baseload demand, but were left with limited options to provide load-following and peaking services.[7] This led utilities to actively evaluate pumped hydro (along with other storage technologies) as alternatives to fossil-fueled intermediate load and peaking units. The economic analysis and justification of new energy storage facilities during this period was based on a direct comparison of the energy and capacity provided by energy storage to an equivalently sized fossil plant, (choosing the lower net-cost option) which largely ignored any additional operational benefits energy storage can provide.[8] Figure 3.1 provides a simple framework of comparing these technologies over time. In the figure, the variable (fuel-related) costs are shown for a storage device and fossil- fueled alternatives. In this figure, the storage technology is assumed to be fueled with off- peak coal and has an effective round-trip efficiency of 75%. [9]

Figure 3.1 shows that the variable cost of providing energy from a storage device was much lower than alternatives available in the mid-1970s and early 1980s. While Figure 3.1 provides the fuel costs, the total economics of storage must also consider the fixed costs. During the mid- to late 1970s, gas-fired combined cycle plants were not significantly less expensive than pumped hydro, with cost estimates of $1 10-$280/kW for a 10-hour PHS device and $ 175-275/kW for a combined-cycle generator (EPRI 1976). As a result, pumped hydro appeared more economic than alternative generation sources during this time period, even without considering additional operation benefits. It was expected that oil and gas prices would remain high, and that off-peak energy would be widely available (and even less expensive) due to anticipated large-scale deployment of nuclear power plants.

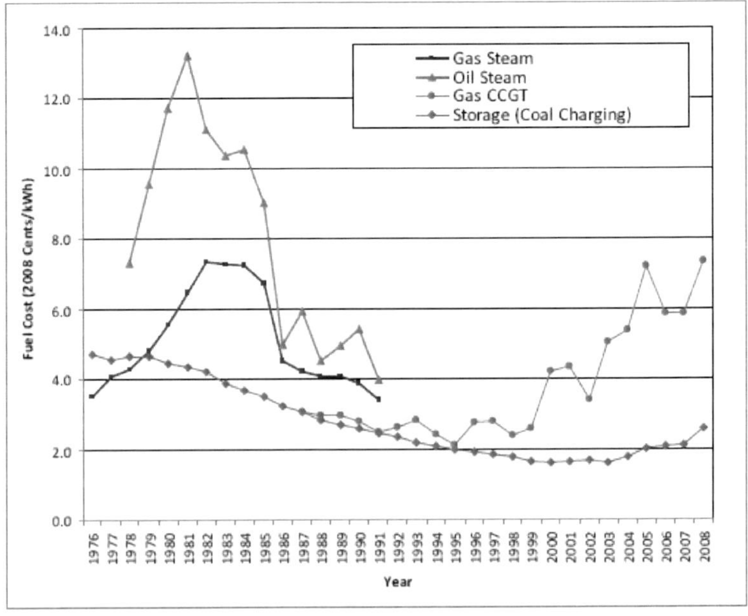

Figure 3.1. Historical fuel costs for intermediate load power plants[10]

During the mid- to late 1970s, much of the nation's 20 GW of pumped hydro storage was initiated (ASCE 1993), along with significant research and development in a variety of other storage technologies including several battery types, capacitors, flywheels, and superconducting magnetic storage (DOE 1977). Growth projections for energy storage during this period included significant increases of several types (Boyd et al. 1983). However, most PHS development, along with interest in and deployment of other emerging storage technologies, ended in the 1 980s after dramatic reduction in the price of natural gas, increased efficiency and reduced costs of flexible combined-cycle and simple-cycle natural gas turbines, and repeal of the Fuel Use Act in 1987. While estimates from the 1970s place combined-cycle gas turbine (CCGT) units and PHS at similar costs, by the early 2000s, PHS was estimated to be about twice the cost of a CCGT. As a result, even with the increased cost of gas, this dramatic increase in PHS cost (along with the many other factors discussed previously) limited the economic competitiveness of PHS vs. gas-fired generators.[11] Furthermore, while coal prices continued to drop, the limited nuclear build-out eliminated a source of low-cost off-peak electricity. Finally, the simplistic treatment of the economic benefits of energy storage technologies was also a limiting factor. One of the main benefits of energy storage is its ability to provide multiple services, including load leveling (and associated benefits such as a reduction in cycling-induced maintenance) along with regulation and contingency reserves and firm capacity.[12] However, it has always been somewhat difficult to quantify these various value streams without fairly sophisticated modeling and simulation methods, (especially before the advent of energy and ancillary service markets, which will be discussed in the next section). Because the economic analysis is difficult, and benefits of storage are often uncertain, utilities tend to rely on more traditional generation assets, especially in regulated utilities where risk is minimized and new technologies are adopted relatively slowly.[13]

Combined, these factors have restricted deployment of utility-scale energy storage in the United States. Besides PHS, deployment has been limited to a single 110 MW compressed-air energy storage (CAES) facility, and a variety of small projects.[14] A more comprehensive discussion of energy storage technologies and their status is provided in Section 5.

3.2. The Economics of Energy Storage in Restructured Markets

Despite the lack of significant new construction, interest in energy storage never completely disappeared during the period of low-cost peaking fuels. Research and development has continued, along with an increasing number of proposed projects.

Recent renewed interest in energy storage has been motivated by at least five factors: advances in storage technologies, an increase in fossil fuel prices, the development of deregulated energy markets including markets for high-value ancillary services, challenges to siting new transmission and distribution facilities, and the perceived need and opportunities for storage with variable renewable generators.

Emergence of wholesale electricity markets along with increased volatility in natural gas prices has created new opportunities and interest in energy storage. As shown in Figure 3.1, rising natural gas prices in the early 2000s increased the cost-competitiveness of energy storage. However, perhaps the single greatest motivation for proposals to build new energy storage is the creation of markets for both energy and ancillary services including regulation, contingency reserves, and capacity. As of 2009, wholesale energy markets exist in parts of more than 30 states and cover about two-thirds of the U.S. population (IRC 2009). The markets provide real, transparent data for both utilities and independent power producers to consider the opportunities for energy storage. Market data allows evaluation of both the economic yield and optimum location of energy storage devices for arbitrage – the ability to purchase low-cost off-peak energy and re-sell this energy during on-peak periods. Furthermore, the benefits of providing operating reserves and other ancillary services from energy storage can now be evaluated. Previously, the value of these services was largely "hidden" in utilities' cost of service, and the cost of providing operating reserves, for example, was rarely calculated. The high value of these services is now recognized, and the advantages of energy storage in providing these services is evident, especially because these services generally require fast response and limited actual energy delivery, two qualities that are well-suited to many energy storage devices.

Historical market data can be used to evaluate the potential profitability of energy storage devices that provide various services. Table 3.1 provides the results of several studies of U.S. electricity markets.

Table 3.1. Historical Values of Energy Storage in Restructured Electricity Markets

Market Evaluated	Location	Years Evaluated	Annual Value ($/kW)	Assumptions
Energy Arbitrage	PJM[a]	2002-2007	$60-$115	12 hour, 80% efficient device. Range of efficiencies and sizes evaluated[15]
	NYISO[b]	2001-2005	$87-$240 (NYC) $29-$84 (rest)	10 hour, 83% efficient device. Range of efficiencies and sizes evaluated.
	USA[c]	1997-2001	$37-$45	80% efficient device, Covers NE, No Cal, PJM
	CA[d]	2003	$49	10 hour, 90% efficient device.
Regulation	NYISO[b]	2001-2005	$163-248	
	USA[e]	2003-2006	$236-$429	PJM, NYISO, ERCOT, ISONE
Contingency Reserves	USA[e]	2004-2005	$66-$149	PJM, NYISO, ERCOT, ISONE

[a] Sioshansi et al. 2009
[b] Walawalkar et al. 2007
[c] Figueiredo et al. 2006
[d] Eyer et al. 2004
[e] Denholm and Letendre 2007

The values in Table 3.1 can be translated into a maximum capital cost for the applicable storage technology (equal to the maximum cost of a storage device that can be supported by the revenues available). Figure 3.2 provides a generic conversion between annual costs and total capital costs. This conversion is performed by dividing the annual revenues by the capital charge rate, which produces a total capital cost.[16] The capital charge rate (also referred to as a fixed-charge rate or capital recovery factor) refers to the fraction of the total capital cost that is paid each year to finance the plant. It should be noted that this cost does not include any operation and maintenance costs.[17]

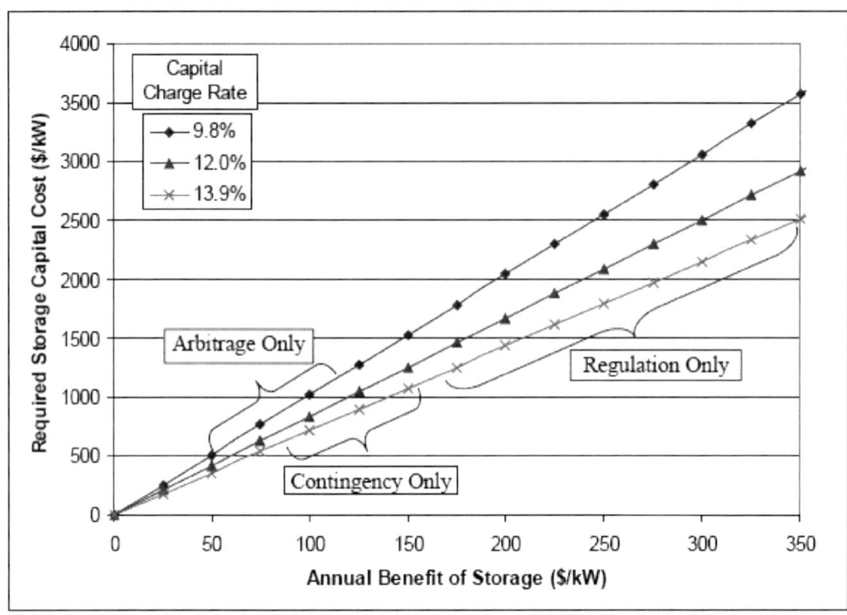

Figure 3.2. Relationship between the annual benefit of storage and capital cost using different capital charge rates[18]

Figure 3.2 shows the range of values and corresponding capital costs for three types of operation: energy arbitrage, contingency reserves, and frequency regulation. In general, energy arbitrage provides the least value. Outside of New York City, the maximum annual value in Table 3.1 for arbitrage was $11 5/kW (for a 20-hour device), which would translate into a capital cost of $827-1,170/kW – this is below current estimates for most energy storage technologies. While there is significant uncertainty in costs, most energy storage assessments indicate that few commercially available bulk energy storage technologies are deployable for less than $ 1,000/kW.[19] This reveals the same challenging economics as comparing a storage device to a conventional generator as discussed in Section 3-1. The value of energy arbitrage alone does not appear to justify the deployment of energy storage at current technology costs and electricity prices.

The value of energy storage increases when taking advantage of other individual sources of revenue or even combined services. A device with sufficient energy capacity for energy arbitrage would likely be able to receive capacity payments in locations where capacity markets now exist; recent data in the PJM market indicates an additional potential value of $40-90/kW-year.[20]

Alternatively, contingency reserves offer a higher value[21] than energy arbitrage and also require less energy capacity. Obtaining the values for energy arbitrage in Table 3.1 generally requires a device with at least 10 hours of storage capacity; contingency reserves can require as little as 30 minutes, depending on the market and market reliability rules (PJM 2009b). The challenge for a device providing contingency reserves is that the device must be able to respond rapidly, typically in a few minutes or less. Frequency regulation is even more demanding, requiring continuous changes in output, frequent cycling, and fast response. It is also the highest-value opportunity for an energy storage device, and has been the focus of many potential energy storage applications, especially given its fairly small energy requirements.[22]

3.3. Other Applications of Energy Storage

In addition to energy arbitrage and operating reserves, there are several other services that energy storage has provided or could provide in the current grid. Several of these applications are discussed below.

Transmission and Distribution

In addition to generation, storage can act as an alternative or supplement to new transmission and distribution (T&D). Distribution systems must be sized for peak demand; as demand grows, new systems (both lines and substations) must be installed, often only to meet the peak demand for a few hours per year. New distribution lines may be difficult or expensive to build, and can be avoided or deferred by deploying distributed storage located near the load (Nourai 2007). (Energy can be stored during off-peak periods when the distribution system is lightly loaded, and discharged during peak periods when the system may otherwise be overloaded.) Energy storage can also reduce the high line-loss rates that occur during peak demand (Nourai et al. 2008).

Black-Start

Black-start provides capacity and energy after a system failure. A black-start unit provides energy to help other units restart and provide a reference frequency for synchronization. Pumped hydro units have been used for this application.[23]

Power Quality and Stability

Energy storage can be used to assist in a general class of services referred to as power quality and stability. Power quality refers to voltage spikes, sags, momentary outages, and harmonics. Storage devices are often used at customer load sites to buffer sensitive equipment against power quality issues. Electric power systems can also experience oscillations of frequency and voltage. Unless damped, these disturbances can limit the ability of utilities to transmit power and affect the stability and reliability of the entire system. System stability requires response times of less than a second, and can be met by a variety of devices including fast-responding energy storage.

End-Use/Remote Applications

Other applications for energy storage are at the end use. Storage can provide firm power for off-grid homes, but also can provide value when grid-tied through management of time-of-use rates, or demand charges in large commercial and industrial buildings. Energy storage also provides emergency and backup power for increased reliability. In many cases, end-use applications have analogous applications in the grid as a whole (and potentially compete with these applications). For example, using energy storage to time- shift end use is functionally equivalent to energy arbitrage, and a flatter load on the demand side reduces the potential need for load-leveling in central storage applications (and vice-versa). To be economic, end-use applications require time-varying prices and are extremely site-specific.

3.4. Summary of Energy Storage Applications in the Current Grid

Table 3.2 provides a summary of the various applications of energy storage commonly discussed in the literature. Each of these applications provides a potential value to a merchant storage operator in a restructured market or a source of cost reduction to the system.

It should be noted that Table 3.2 does not include any dedicated renewables applications. Historical motivations for energy storage deployment were based on the challenges of meeting variations in demand using conventional thermal generators. Much of the current attention for energy storage is based on its potential application with renewable energy (primarily solar and wind). Energy storage is seen as a means to “firm” or “shape” the output from variable renewable generators. However, the actual need for storage to perform these roles has yet to be quantified. In addition, it is unclear whether the electric power industry must create entirely new "classes" of energy and capacity services to deal with the increased uncertainty and variability created by large-scale deployment of variable renewables. Determining the role of energy storage with renewables first requires examining the impacts of variable generators on the grid and how these impacts may require the use of various enabling technologies.

Table 3.2. Traditional Major Grid Applications of Energy Storage

Application	Description	System Benefits when Provided by Storage	Timescale of Operation
Load Leveling/ Arbitrage	Purchasing low-cost off-peak energy and selling it during periods of high prices.	Increases utilization of baseload power plants and decrease use of peaking plants. Can lower system fuel costs, and potentially reduce emissions if peaking units have low efficiency.	Response in minutes to hours. Discharge time of hours.
Firm Capacity	Provide reliable capacity to meet peak system demand.	Replace (or function as) peaking generators.	Must be able to discharge continuously for several hours or more.
Operating Reserves			
Regulation	Fast responding increase or decrease in generation (or load) to respond to random, unpredictable variations in demand.	Reduces use of partially loaded thermal generators, potentially reducing both fuel use and emissions.	Unit must be able to respond in seconds to minutes. Discharge time is typically minutes. Service is theoretically "net zero" energy over extended time periods.
Contingency Spinning Reserve[24]	Fast response increase in generation (or decrease load) to respond to a contingency such as a generator failure.	Same as regulation.	Unit must begin responding immediately and be fully responsive within 10 minutes. Must be able to hold output for 30 minutes to 2 hours depending on the market. Service is infrequently called.[25]

Table 3.2. (Continued)

Application	Description	System Benefits when Provided by Storage	Timescale of Operation
Replacement/ Supplemental	Units brought on-line to replace spinning units.	Limited. Replacement reserve is typically a low-value service	Typical response time requirement of 30-60 minutes depending on market minutes. Discharge time may be several hours.
Ramping/Load Following	Follow longer term (hourly) changes in electricity demand.	Reduces use of partially loaded thermal generators, potentially reducing both fuel use and emissions. Price is "embedded" in existing energy markets, but not explicitly valued, so somewhat difficult to capture.	Response time in minutes to hours. Discharge time may be minutes to hours.
T&D Replacement and Deferral	Reduce loading on T&D system during peak times.	Provides an alternative to expensive and potentially difficult to site transmission and distribution lines and substations. Distribution deferral is not captured in existing markets.	Response in minutes to hours. Discharge time of hours.
Black-Start	Units brought online to start system after a system-wide failure (blackout).	Limited. May replace conventional generators such as combustion turbines or diesel generators.	Response time requirement is several minutes to over an hour. Discharge time requirement may be several to many hours.[26]
End-Use Applications			
TOU Rates	Functionally the same as arbitrage, just at the customer site.	Same as arbitrage.	Same as arbitrage.
Demand Charge Reduction	Functionally the same as firm capacity, just at the customer site.	Same as firm capacity.	Same as firm capacity.
Backup Power/ UPS/Power Quality	Functionally the same as contingency reserve, just at the customer site.	Benefits are primarily to the customer.	Instantaneous response. Discharge time depends on level of reliability needed by customer.

4. IMPACTS OF RENEWABLES ON THE GRID AND THE ROLE OF ENABLING TECHNOLOGIES

The introduction of variable renewables is now one of the primary drivers behind renewed interest in energy storage. A common claim is that renewables such as wind and solar are intermittent and unreliable, and require backup and firming to be useful in a utility system – energy produced by wind and solar should be "smoothed" or shifted to times when the wind is not blowing or the sun is not shining using energy storage. These statements are generally qualitative in nature and provide little insight into the actual role of renewables in the grid, (including their costs and benefits) or the potential use of energy storage or other enabling technologies.

To evaluate the actual role of energy storage in a grid with large amounts of variable renewable generation, we must first return to our previous discussion of the variability of electric demand, and how the conventional generators currently meet this demand. As discussed in Section 2, tremendous variation in daily demand is met by the constant up- and-down cycling of generators. In addition to this daily cycling, frequency regulation and contingency reserves are provided by partly loaded generators and responsive load.[27] Most of these "flexible" generators are hydro units, combustion turbines, some combined-cycle plants, and even large thermal generators, as well as the existing PHS.

Variable generation (VG) [28] will change how the existing power plant mix is operated, because its output is unlike conventional dispatchable generators. It is easiest to understand the impact of VG technologies on the grid by considering them as a source of demand reduction with unique temporal characteristics. Instead of considering wind or PV as a source of generation, they can be considered a reduction in load with conventional generators meeting the "residual load" of normal demand minus the electricity produced by renewable generators.

Figure 4.1 illustrates this framework for understanding the impacts of variable renewables. In this figure, renewable generation is subtracted from the normal load, showing the "residual" or net load that the utility would need to meet with conventional sources. [29] The benefits to the utility include reduced fuel use (and associated emissions)[30] and a somewhat reduced need for overall system capacity (this is relatively small for wind but can be significant for solar given its coincidence with load.)[31] There are also four significant impacts that change how the system must be operated and affect costs.

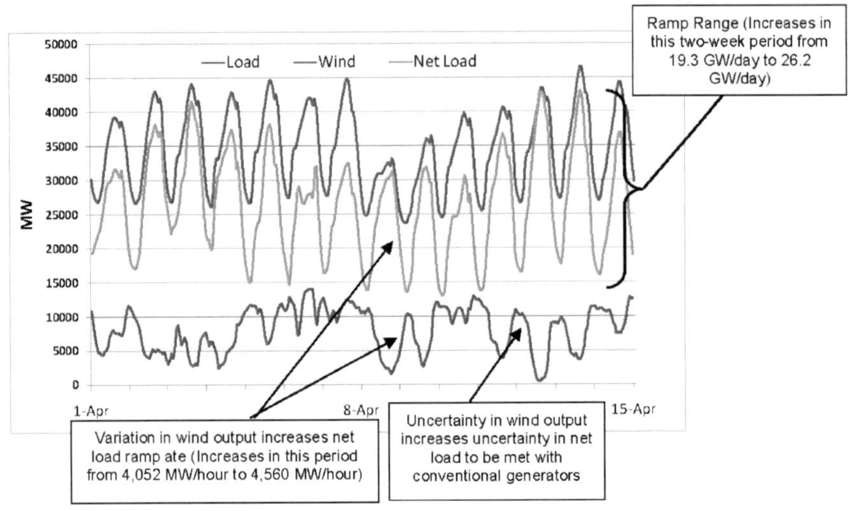

Figure 4.1. Impact of net load from increased use of renewable energy

First is the increased need for frequency regulation, because wind can increase the short-term variability of the net load (not illustrated on the chart).[32] Second is the increase in the ramping rate, or the speed at which load-following units must increase and decrease output. The third impact is the uncertainty in the wind resource and resulting net load.[33] The final impact is the increase in overall ramping range – the difference between the daily minimum and maximum demand – and the associated reduction in minimum load, which can force baseload generators to reduce output; and in extreme cases, force the units to cycle off during periods of high wind output. Together, the increased variability of the net load requires a greater amount of flexibility and operating reserves in the system, with more ramping capability to meet both the predicted and unpredicted variability. The use of these variable and uncertain resources will require changes in the operation of the remaining system, and this will incur additional costs, typically referred to as integration costs.

4.1. Costs of Wind and Solar Integration from Previous Studies

Concerns about grid reliability and the cost impacts of wind have driven a large number of wind integration studies. These studies use utility simulation tools[34] and statistical analysis to model systems with and without wind and calculate the integration costs of wind.

Table 4.1. Summary of Recent Wind Integration Cost Studies (DeCesaro et al. 2009)

Date	Study	Wind Capacity Penetration (%)	Regulation Cost ($/MWh)	Load-Following Cost ($/MWh)	Unit Commitment Cost ($/MWh)	Other ($/MWh)	Tot Oper. Cost Impact ($/MWh)
2003	Xcel-UWIG	3.5	0	0.41	1.44	Na	**1.85**
2003	WE Energies	29	1.02	0.15	1.75	Na	**2.92**
2004	Xcel-MNDOC	15	0.23	na	4.37	Na	**4.6**
2005	PacifiCorp-2004	11	0	1.48	3.16	Na	**4.64**
2006	Calif. (multi-year)[a]	4	0.45	trace	trace	Na	**0.45**
2006	Xcel-PSCo[b]	15	0.2	na	3.32	1.45	**4.97**
2006	MN-MISO[c]	36	na	na	na	na	**4.41**
2007	Puget Sound Energy	12	na	na	na	na	**6.94**
2007	Arizona Pub. Service	15	0.37	2.65	1.06	na	**4.08**
2007	Avista Utilities[d]	30	1.43	4.4	3	na	**8.84**
2007	Idaho Power	20	na	na	na	na	**7.92**
2007	PacifiCorp-2007	18	na	1.1	4	na	**5.1**
2008	Xcel-PSCo[e]	20	na	na	na	na	**8.56**

[a] Regulation costs represent 3-year average.

[b] The Xcel/PSCO study also examine the cost of gas supply scheduling. Wind increases the uncertainty of gas requirements and may increase costs of gas supply contracts.

[c] Highest over 3-year evaluation period. 30.7% capacity penetration corresponding to 25% energy penetration

[d] Unit commitment includes cost of wind forecast error.

[e] This integration cost reflects a $10/MMBtu natural gas scenario. This cost is much higher than the integration cost calculated for Xcel-PSCo in 2006, in large measure due to the higher natural gas price: had the gas price from the 2006 study been used in the 2008 study, the integration cost would drop from $8.56/MWh to $5.13/MWh.

The basic methodology behind these studies is to compare a base case without wind to a case with wind, evaluating technical impacts and costs. The studies calculate the additional costs of adding operating reserves as well as the other system changes needed to reliably address the increased uncertainty and variability associated with wind generation.

Table 4.1 provides examples of several integration studies from various parts of the United States. In these studies, integration costs are typically divided into three types, based on the timescales important to reliable and economic power system operation. These three types are the first three of the four impacts discussed previously:

Regulation – the increased costs that result from providing short-term ramping (seconds to minutes) resulting from wind deployment.

Load following – the increased costs that result from providing the hourly ramping requirements resulting from wind deployment.

Wind uncertainty – the increased costs that result from having a suboptimal mix of units online because of errors in the wind forecast. This is typically called unit commitment or scheduling cost because it involves costs associated with committing (turning on) too few or too many slow-starting, but lower operational-cost units than would have been committed if the wind forecast been more accurate.

The overall cost impact of accommodating wind variability in these studies is typically less than $5/MWh (0.5 cents/kWh), adding less than 10% to the cost of wind energy.[35] The majority of these costs appear to be a result of wind forecasting errors and uncertainty – resulting in "unit commitment" errors where too little or too much capacity is kept online. It is worth noting that wind forecasting is an area of active research, and these errors are expected to decrease in time, which could potentially lead to a corresponding decrease in unit commitment errors and associated costs.

The explanation for this relatively modest impact on costs is largely based on the already significant variation in normal load. The large amount of flexible generation already available to meet the variability in demand has the ability to respond to the greater variability caused by the large-scale deployment of wind. Furthermore, these studies have found significant benefits of spatial diversity – just because the wind isn't blowing in one location, it may be in another. The combination of multiple wind sites tends to smooth out the aggregated wind generation in a system, which reduces the per-unit size of ramps and mitigates the range of flexibility required.[36]

Far less work has been performed on the operational impacts of large-scale solar generation due largely to lower deployment rates compared to wind.[37] In addition, there is insufficient solar data available to estimate impacts on frequency regulation and ramping (Lew et al. 2009.) One study of PV on the Xcel Colorado utility system found integration costs of between $3.5 1/MWh and $7.14/MWh for a scenario examining 800 MW of solar in a 6,922 MW peaking system, with gas prices ranging from $7.83 to $11. 83/MMBTU (EnerNex 2009). Additional studies are ongoing, but it will be some time until knowledge of solar's impact on the grid and associated costs are understood to the degree of wind.[38]

In reality, while these studies divide the costs into three main categories, the source of actual integration costs is largely associated with the fuel costs needed to provide the additional required reserves, along with some variable operations and maintenance costs. To provide regulation and load following, the additional variability requires that utilities run more flexible generators (such as gas-fired units instead of coal units, or simple-cycle turbines instead of combined-cycle turbines) to ensure that additional ramping requirements can be met. This was illustrated previously in Figure 2.3 where a "nonoptimal" mix of generators was needed to provide the ability to adjust output in response to contingencies and other variations in demand. This combination of higher fuel costs and lower-efficiency units results in increased cost of fuel per unit of electricity generated as opposed to the "no-wind" cases.[39] Additional fuel costs occur from keeping units at part-load, ready to respond to the variability, or from more frequent unit starts. The costs associated with unit commitment or scheduling are also largely captured in increased fuel costs. Most large thermal generators must be scheduled several hours (or even days) in advance to be ready when needed. Ideally, utilities schedule and operate only as many plants as needed to meet energy and reserve requirements at each moment in time. If utilities over-schedule (turn on too many plants), they will have many plants running at part-load, and have incurred higher than needed start-up costs. So, if they under-predict the wind, they will commit and start up too many plants, which incurs greater fuel (and other) costs than needed if the wind and corresponding net load had been forecasted accurately. Conversely, if the wind forecast is too high and the net load is higher than expected, insufficient thermal generation may be committed to cover the unexpected shortfall in capacity. A worst-case scenario would be a partial blackout; but the likely result is the use of high-cost "quick-start" units in real-time, purchasing expensive energy from neighboring utilities (if available), or paying customers a premium to curtail load – all while lower cost units are sitting idle. An example of this is the ERCOT event of Feb. 26, 2008 (Ela and Kirby 2008). On this date, a combination of events – including a greater than predicted demand for energy, a forced outage of a conventional unit, the wind forecast not being given to the system operators, and a lower than expected wind output – resulted in too little capacity online to meet load. As a result, the ERCOT system needed to deploy high-cost quick-start units, as well as pay customers to curtail load through its "load acting as a resource" program.[40] All of this occurred while lower-cost units were idle because the combination of events was unanticipated.[41]

While the bulk of the costs associated with wind integration are due to fuel use, the increased cycling also increases wear and tear on generators, which imposes extra maintenance costs.[42]

Results from wind integration studies almost universally come to the conclusion that at the penetrations studied to date (up to about 30% on an energy basis), the analyzed systems do not need additional energy storage to accommodate wind's variability and maintain reliable service.[43] In the studied systems, no new generation technologies are required; but there are some potentially significant operational changes needed to maintain the present level of reliability (along with significant transmission additions needed to exchange resources over larger areas).[44] However, they do not necessarily find the "cost optimal" solution, which may include energy storage or some alternative mix of generation to further reduce the cost of wind integration. Furthermore, these studies have not evaluated much higher penetration levels of RE, where additional system constraints may require additional enabling technologies such as energy storage.

4.2. Limiting Factors for Integration of Wind and Solar Energy

To date, integration studies in the United States have found that variable generation sources can be incorporated into the grid by changing operational practices to address the increased ramping requirements over various timescales. At higher penetrations (beyond those already studied), the required ramp ranges will increase, which adds additional costs and the need for fast-responding generation resources. However, there are additional constraints on the system that will present additional challenges. These constraints are based on the simple coincidence of renewable energy supply and demand for electricity, combined with the operational limits on generators providing baseload power and operating reserves. Of the four operational cost impacts listed in the beginning of Section 4 (regulation, load following, scheduling, and ramping range), only the first three present major quantifiable costs in

U.S. studies as of the end of 2009. Yet, it is the fourth constraint that may present an economic upper limit on variable renewable penetration without the use of enabling technologies.[45]

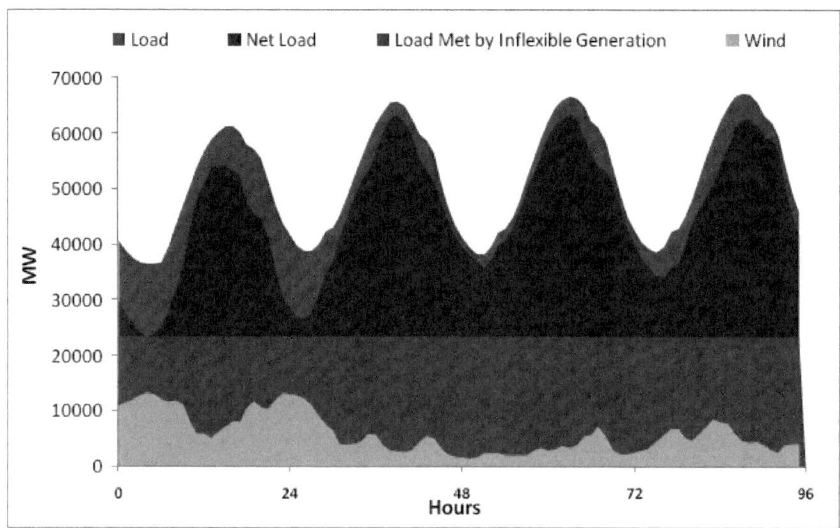

Figure 4.2. Dispatch with low VG penetration (wind providing 8.5% of load)

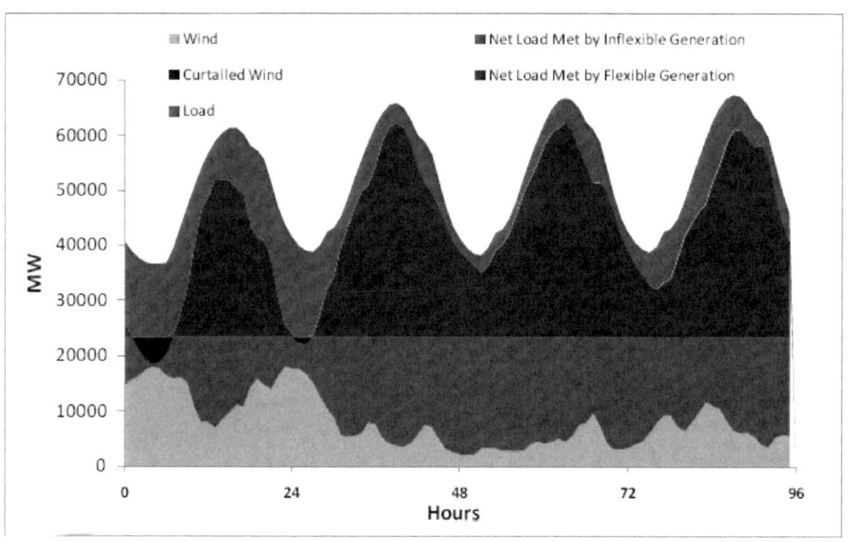

Figure 4.3. Dispatch with higher VG penetration (wind providing 16% of load)

As discussed in Section 1, in current electric power systems, electricity is generated by two general types of generators: baseload generators, which run at nearly constant output; and load-following units (including both intermediate load and peaking plants), which meet the variation in demand as well as provide operating reserves. At current penetrations of wind and solar in the United States, and at the levels studied in most integration studies, wind and solar generation primarily displaces flexible load-following generators. Figures 4.2 and 4.3 illustrate this issue by providing the impacts of increasing amounts of wind generation in a simulated grid. [46] In Figure 4.2, wind provides 8.5 % of the energy in this four-day period and displaces the output from a mix of gas-fired units, which are already typically used to follow load. Because these generators are designed to vary output, they can do this with modest cost penalties as analyzed in the wind integration studies discussed previously.

At higher penetration of RE, the ability of conventional generators to reduce output becomes an increasing concern. In Figure 4.3, wind now provides 16% of the total demand in this four-day period. Variable renewables begin to displace units that are traditionally not cycled, and the ability of these thermal generators to reduce output may become constrained. If the baseload generators cannot reduce output (and some other use cannot be found for this "excess generation"),[47] then wind energy will need to be curtailed – this occurs in the overnight periods in the first two days of this scenario.

Utilities in the United States have expressed concern about their systems "bottoming out" due to the minimum generation requirements during overnight hours, and being unable to accommodate more variable generation during these periods. Minimum generation constraints (and resulting wind curtailment) are already a real occurrence in the Danish power system, which has a large installed base of wind generation (Ackermann et al. 2009). Due to its reliance on combined heat and power electricity plants for district heating, the Danish system needs to keep many of its power plants running for heat. Large demand for heat sometimes occurs during cold, windy evenings, when electricity demand is low and wind generation is high. This combination sometimes results in an oversupply of generation, which forces curtailment of wind energy production. The need to curtail wind due to minimum load constraints has also been identified as an important component of future power systems in the United States.[48] Modern wind turbines can reliably curtail output, but this is largely undesirable because curtailment throws away cost-free and emissions-free energy.[49]

The actual minimum load is a function of several factors including the mix of conventional generation, as well as the amount of reserves and the types of generators providing those reserves. The ability to cycle conventional units is both a technical and economic issue – there are technical limits to how much power plants of all types can be turned down. Large coal plants are often restricted to operating in the range of 50-100% of full capacity, but there is significant uncertainty about this limit. This is partly because utilities often have limited experiencc with cycling large coal plants, and have expressed concern about potentially excessive maintenance impacts[50] and even safety. It should be noted that because cycling costs are not universally captured in operational models, they may be ignored or underestimated in wind and solar integration studies.[51] The impact of VG on power plant cycling is an active area of research, especially considering the evolving grid and introduction of more responsive generation. [52]

It is unclear what changes in operation practices or generator modification utilities will need to make to cycle below their current minimum load points, which now typically occur during the early morning in spring. In a number of markets, energy prices have dropped below the actual variable (fuel) cost of producing electricity on a number of occasions. This indicates that power plant operators are willing to sell energy at a loss to avoid further reducing output. Figure 4.4 provides one example in the PJM market in 2002, where the price of electricity fell below the variable cost of generation (indicated by the dotted line)[53] from coal-fired units for about 100 hours during periods near the annual minimum. While not definitive, this indicates that under current operational practices, utilities in some systems may be uncomfortable or unable to cycle much below current minimum load levels.[54] In other locations, there may be a greater operating range if plant operators become more comfortable with cycling individual units. As an example, a recent wind integration study of the existing ERCOT system suggested the capability of the system to cycle down to a net load of about 13 GW, compared to the recent annual minimum loads of about 20 GW. However, this would require the coal fleet to cycle to below 50% of rated capacity, and gas plants to perform over an even greater cycling range (GE Energy 2008).

Overall, the ability to accommodate a variable and uncertain net load has been described as a system's flexibility. System flexibility varies by system and over time as new technologies are developed and power plants are retired. In addition, VG deployment in the United States is still relatively small, and utilities have yet to evaluate the true cycling limits on conventional generators and their associated costs. Also, the additional reserve requirements due to VG at high penetration are still uncertain. As a result, it is not possible to precisely estimate the costs of VG integration or the amount of curtailment at very high penetration; it is also difficult to define with certainty the value of energy storage or other enabling technologies. It is clear, however, that substantial increase in the penetration of wind energy without storage will require changes in grid operation to reduce curtailment.

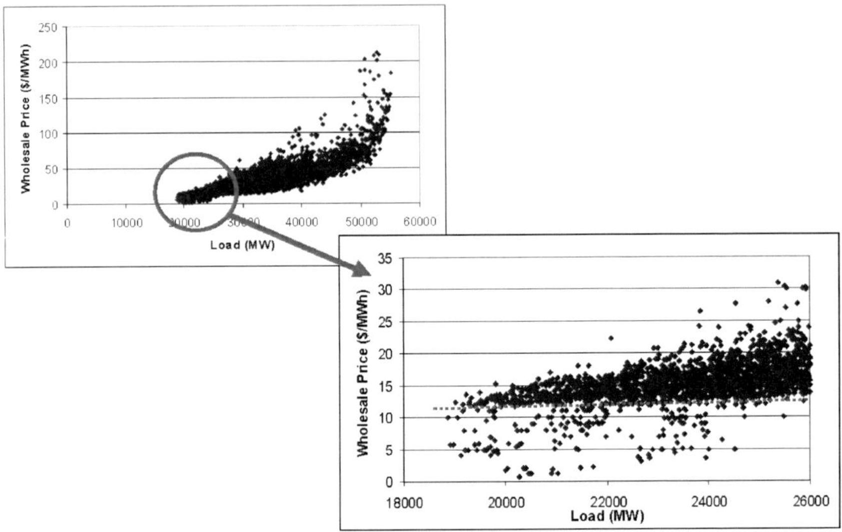

Figure 4.4. Relationship between price and load in PJM in 2002

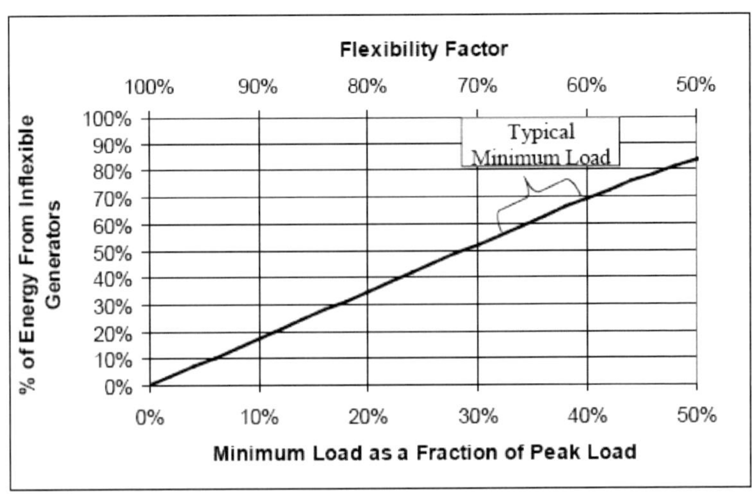

Figure 4.5. Contribution from inflexible generation

Figure 4.5 illustrates the fraction of a system's peak load derived from inflexible units as a function of the minimum load point. This is also expressed more generally as the system's "flexibility factor," which is defined as the fraction below the annual peak to which conventional generators can cycle (Denholm and Margolis 2007a and 2007b). In the current grid, the annual minimum load point is typically 30%-40% of annual peak load (or a flexibility factor of 60%-70%). If units cannot be cycled below this point, they will provide 55%-70% of a system's energy; and even with storage, this level of inflexibility leaves only 30%-40% of a system's energy for variable resources. One of the major conclusions of wind integration studies looking at higher penetrations is that minimum load points will need to be lowered substantially below their current annual minimums.[55]

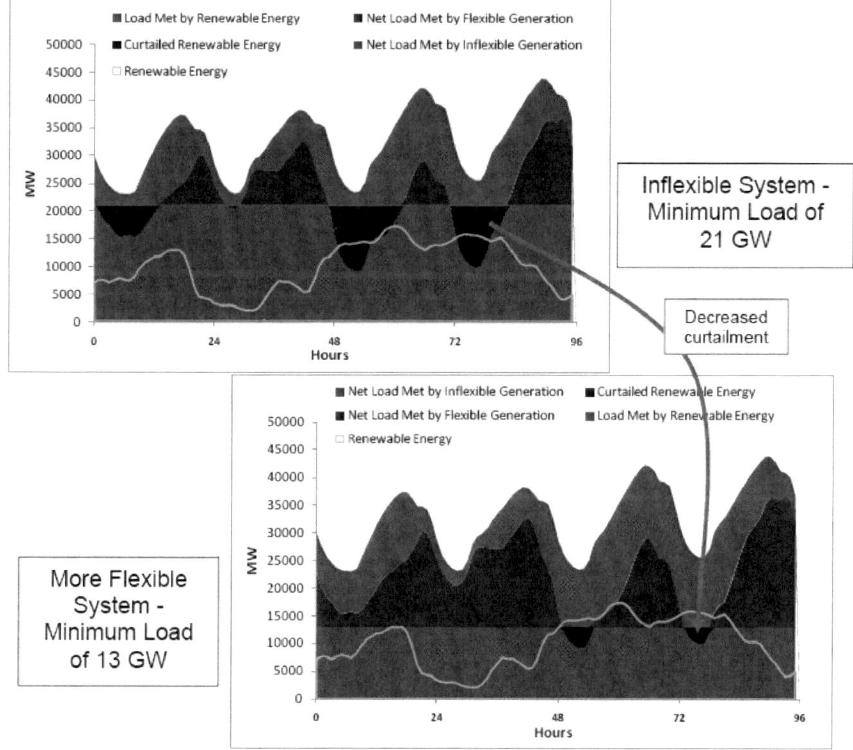

Figure 4.6. Effect of decreasing minimum load point on increased use of RE

Alternatively, it is possible to evaluate the relationship between system flexibility and large-scale penetration of VG, both with and without energy storage or other enabling technologies. Figure 4.6 shows an example of how flexibility affects the potential curtailment of wind and solar energy when deployed without storage. These two charts superimpose load data in ERCOT from 2005 with a spatially diverse set of simulated wind and solar data from the same year. [56,57] The simulation places 19 GW of wind and 11 GW of solar (representing an 80%/20% mix of wind and solar on an energy basis) into the 2005 ERCOT system, which had a peak demand of 60.3 GW. In the left chart, the system is assumed to be unable to cycle below the 2005 minimum point of 21 GW, resulting in substantial VG curtailment. In this simulation, wind and solar provide 20% of the grid's annual energy, and 21% of the total renewable energy production is curtailed. The right graph shows the result of increasing flexibility, allowing for a minimum load point of 13 GW. Curtailment has been reduced to less than 3%, and the same amount of variable renewables now provides about 25% of the system's annual energy.

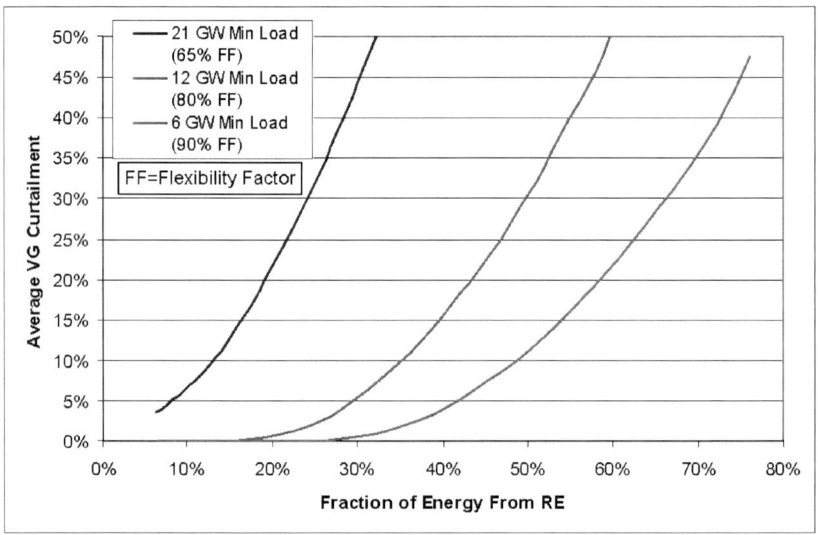

Figure 4.7. Average curtailment rate as a function of VG penetration for different flexibilities in ERCOT

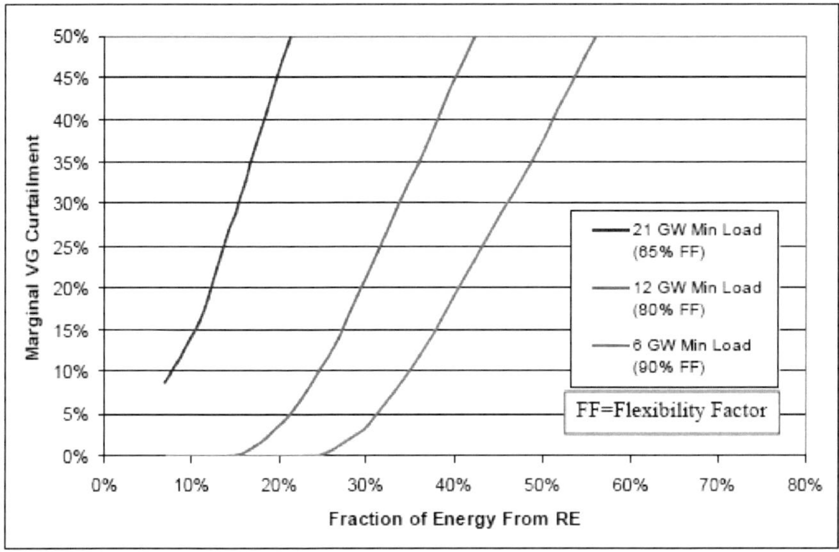

Figure 4.8. Marginal curtailment rate as a function of VG penetration for different system flexibilities in ERCOT (marginal curtailment is defined as the percentage curtailed as a function of the incremental penetration)

These results can be evaluated more generally by examining the amount of curtailment that will result without energy storage as a function of system flexibility. Figure 4.7 shows example results, with three different minimum load/flexibility factors. The first curve illustrates the curtailment that would result if the system could not cycle below its 2005 annual minimum load of 21 GW. The second curve uses a minimum load of 12 GW, which corresponds roughly to the assumptions used in the 2008 ERCOT wind study (GE 2008). The third curve corresponds to a 6 GW minimum load, which largely eliminates baseload units from the generation mix, replacing them with more flexible units.[58]

Figure 4.7 somewhat obscures the fact that at the margin, curtailment rates can be very high. Figure 4.8 illustrates the marginal curtailment rate, or the curtailment rate of each incremental unit of VG installed in the system. For example, in the 12 GW minimum load curve, the average curtailment rate (Figure 4.7) when VG is providing 25% of the system's electricity demand is less than 3%, which means that less than 3% of all the VG at this penetration level is curtailed. However, the last unit of energy installed to get to this 25% point has a curtailment rate of more than 10%, as illustrated in Figure 4.8.

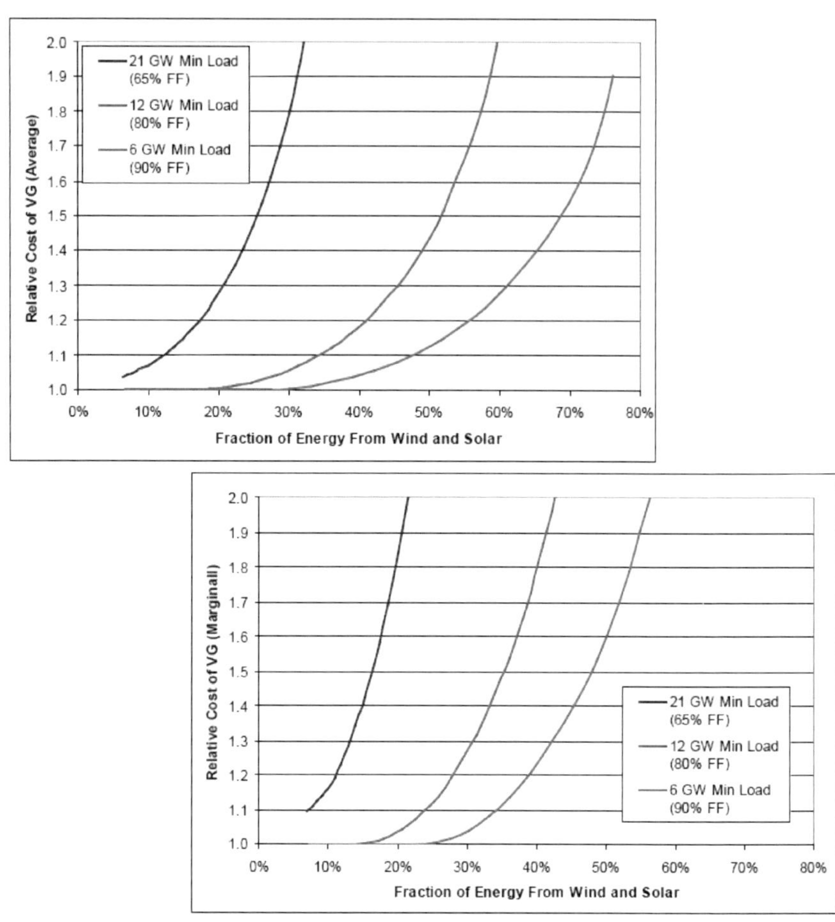

Figure 4.9. Relative cost of VG – average (top chart) and marginal (bottom chart) – as a function of VG penetration for different system flexibilities in ERCOT

Curtailment increases the cost of the VG that is actually used, because curtailment reduces the net capacity factor of the wind and solar generators. Figure 4.9 shows how the effective cost of VG increases due to curtailment. Costs are illustrated in relative terms – a generator with no curtailment has a base cost of 1, which increases with a scale factor equal to (1/1-curtailment rate).

This example can be compared to the results of the previous U.S. wind integration studies that find very little curtailment up to about 20% on an energy basis, due to the existing grid flexibility. Beyond this level (up to about 30%), curtailment increases without significant operational changes (Corbus et al. 2009). In the United States, penetrations of VG beyond 30% have yet to be extensively studied; however, the examples in this section suggest curtailment rates will continue to rise substantially without a significant increase in system flexibility or deployment of other enabling technologies such as energy storage.

The results in Figure 4.9 apply to only a specific mix of VG – and a single system – so they cannot be applied generally. However, they do illustrate the trends that may limit the contribution of VG without enabling technologies. Ultimately, the cost of curtailment must be compared to alternatives such as the cost of storage or other enabling technologies, illustrated in Figure 4.10 and discussed in more detail in Section 5.[59]

Storage provides one solution to avoiding curtailment by absorbing otherwise unusable generation and moving it to times of high net system load (where net load is defined as normal load minus VG). The additional flexibility storage provides is also important. Storage can provide operating reserves, which reduces the need for partially loaded thermal generators that may restrict the contribution of VG. Finally, by providing firm capacity and energy derived from VG sources, storage can effectively replace baseload generation, which reduces the minimum loading limitations.

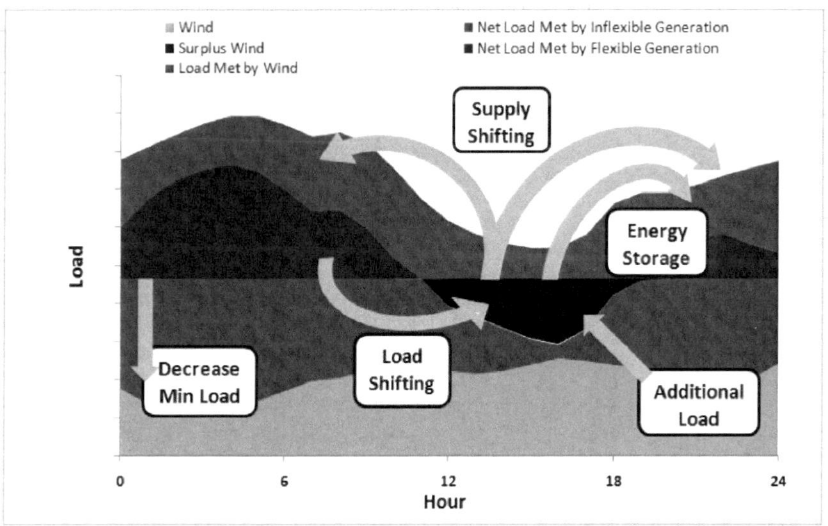

Figure 4.10. Option for increasing the use of VG by decreasing curtailment

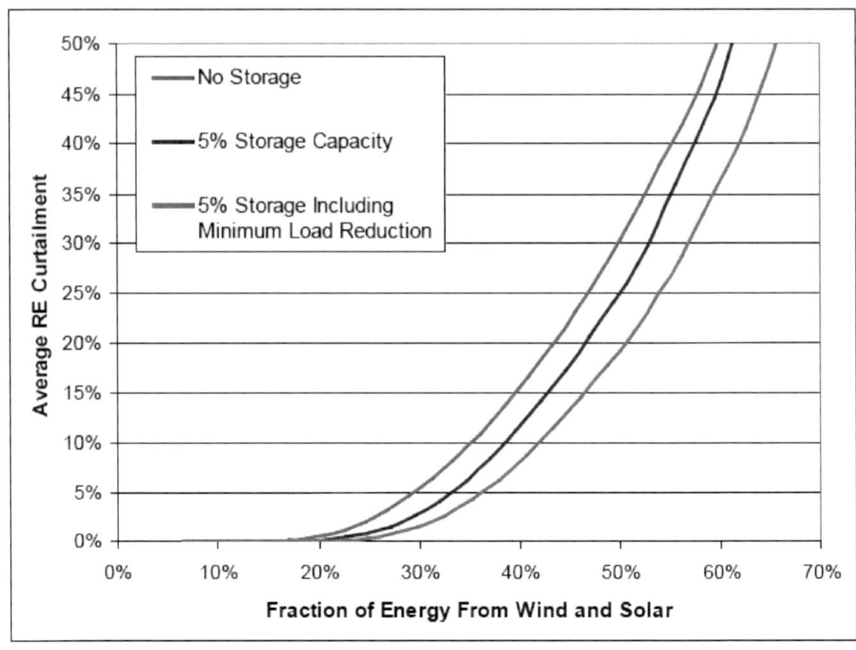

Figure 4.11. Reduction of curtailment resulting from addition of energy storage

Figure 4.11 illustrates how the curtailment rate can be reduced by introducing energy storage.[60] The base case (no storage) is identical to Figure 4.7 with a 12 GW minimum load (or an 80% flexibility factor). In addition, a storage device with 5% of the system's power capacity (or 3 GW in a 60 GW peaking system), 20 hours of energy capacity, and a 75% round-trip efficiency is introduced. If this device is used only to absorb otherwise unusable energy, the curtailment rate is reduced substantially. For example, when VG is providing 50% of the system's energy, about 30% of the VG is curtailed without storage, and about 25% with the 3 GW of storage. This includes the additional losses that occur in the storage process. The device's ability to replace firm capacity and potentially reduce the minimum load constraint further reduces curtailment. Adding the ability to reduce the minimum load by 50% of the device's capacity (or about 1.5 GW) reduces curtailment further – in the 50% VG case, the curtailment rate now drops from 30% without storage to 20% of the total VG.[61]

While storage provides one solution to the mismatch of VG supply and normal demand, there are a variety of options for increasing the use of VG in the grid. Any evaluation of energy storage should consider the many alternative technologies that can increase grid flexibility and enable VG renewables.

5. Storage and Flexibility Options for Renewables-Driven Grid Applications

The previous section indicates that at high penetration of VG, fundamental changes to the grid may be required to accommodate the increased variability of net load and the limited coincidence of VG supply and normal electricity demand. A number of techniques and technologies – described as flexibility resources – have been proposed to accommodate the impacts of VG and ensure the generation mix matches the net-load requirement.

5.1. The Flexibility Supply Curve

Energy storage is one of many technologies proposed to increase grid flexibility and enable greater use of VG. This set of technologies has been described in terms of a flexibility supply curve that can provide responsive energy over various timescales. The flexibility supply curve is conceptually similar to other resource supply curves where (ideally) the lowest-cost resources are used until they are exhausted, then the next (higher-cost) resource is deployed. The analysis in Section 4, for example, was restricted to ERCOT, and did not consider the opportunity to exchange wind and solar energy with surrounding areas by building interconnections with the other U.S. grids. It also did not consider the ability to shift load by incentivizing customers to use less electricity when VG output is low. Overall, utilities have many "flexibility" options for incorporating greater amounts of VG into the grid, many of which may cost less than using energy storage. Figure 5.1 provides a conceptual supply curve including some of these options.

Overall, there are two general "types" of flexibility required by variable sources and offered by technologies in this curve. The first can be described as ramping flexibility, or the ability to follow the variation in net load (in the second-to-minute timescale needed for frequency regulation, or the minutes-to-hours timescale needed for load following and forecast error.) This flexibility is the primary requirement at low penetration, as discussed in Section 4.2. The second type of flexibility is energy flexibility, or the ability to increase the coincidence of VG supply with demand for electricity services, which is described in Section 4.3. A description of several of the sources of flexibility is provided below.

a) **Supply and Reserve Sharing.** This includes the sharing of renewable and conventional supply, operating reserves, and net loads through markets or other mechanisms that effectively increase the area over which supply and demand is balanced. [63] Greater aggregation of loads and reserves has historically been one of the least-cost methods of dealing with demand variability, especially because it often requires operational changes and relatively little new physical infrastructure.[64] It may also require transmission development to increase the spatial diversity of VR resources.
b) **Flexible Generation.** This includes deploying new, more flexible conventional generators as well as increasing flexibility of existing generators. This can be accomplished by modifying equipment and operational practices [65] to increase the load-following, ramping rate, and ramping range of the grid. This also includes introducing new generators that can be brought online quickly to respond to forecast errors.[66] This may also require increased use of natural gas storage to increase use of flexible gas turbines and decrease contractual penalties for forecast errors in natural gas use (Zavadil 2006). Another source of flexible generation is improved use of existing storage and hydro assets.

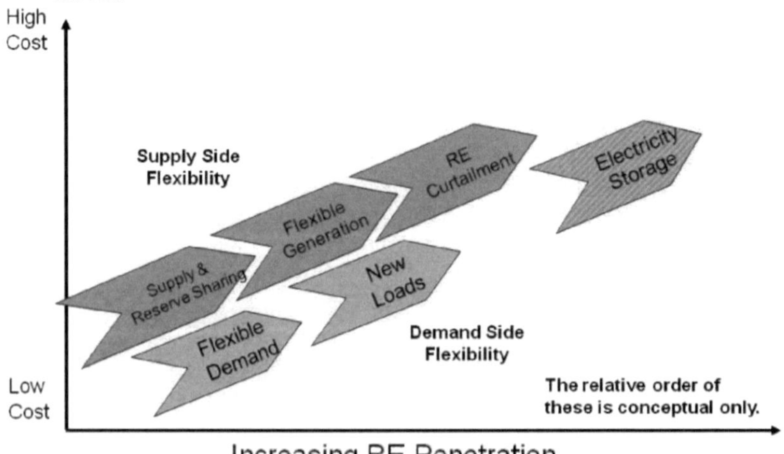

Figure 5.1. Flexibility supply curve[62]

c) **Demand Flexibility.** This includes introducing market or other mechanisms to allow a greater fraction of the load to respond to price variations and provide ancillary services. Responsive demand can provide flexibility over multiple timescales by curtailing demand for short periods or shifting load over several hours. Many of these technologies and processes have been described in terms of a "smart grid" and will require regulatory and policy changes in addition to new technologies. In many locations in the United States, demand is increasingly used as a source of grid services.[67]

d) **VG Curtailment.** Overbuilding VG may result in curtailment of low-value springtime generation, but would allow for a greater overall VG contribution. (This is functionally equivalent to cycling baseload generators in the spring.) Furthermore, curtailed VG provides additional benefits, because it provides a source of operating reserves and potentially allows for de-commitment of thermal units that typically provide these services.

e) **New Loads.** New controllable loads can be added to absorb otherwise unusable VG. Examples include space and process heating, which currently use fossil fuels. Another possibility is fuel production such as hydrogen via electrolysis or shale oil heating. Electrification of transportation using electric vehicles or plug- in hybrid vehicles is also a potential large-scale application. This may also include electric vehicles providing regulation and contingency reserves with or without the use of vehicle to grid (V2G).

f) **Electricity Storage.** Electricity storage encompasses a large number of technologies discussed in Section 5.2

The general classes of flexibility resources listed above represent dozens or even hundreds of individual technologies, each with a potential contribution to increasing grid flexibility. The cost and availability of many flexibility resources has yet to be quantified, and there are a variety of regulatory barriers to completely deploying many flexibility options such as demand response.

The cost of storage needs to be compared to the alternatives – this includes the efficiency losses in the storage process that may be avoided by using other enabling technologies. There are, of course, limits to each option on the curve; and the benefits of spatial diversity and demand response are limited because the VG supply cannot be expected to exactly match the demand for electricity services on the multiple timescales. Because of this, electricity storage is considered a potentially important step on the flexibility supply curve when lower-cost options are saturated or otherwise unavailable.

5.2. Deployment and Operational Balancing of Renewable Energy – Individual Plant Storage vs. Power System Storage

The previous section suggests that incorporating increased levels of VG most efficiently will require a variety of flexibility options. Likewise, the most efficient operation and location of flexibility options – including storage – must be considered. While flexibility resources can be considered renewable enabling technologies, their historical application has been to benefit the grid as a whole. It is often suggested that energy storage be co- located with, and operationally tied to, the output of individual VG facilities. However, this is not how the current system balances the large variability in net demand.

Furthermore, the services needed to address variability and uncertainty are generally the same services that storage currently provides to the grid. Table 5.1 lists several renewable-specific applications that have been proposed (EPRI 2004). As an example, one potential renewable-specific application of storage is "time-shifting" of wind from periods of low demand to periods of high demand. However, this application is fundamentally the same as energy arbitrage, and the benefits of this application are greatest when the energy storage operator can choose from all of the generators in a system, and store energy when the cost is lowest, instead of storing only wind generation.

Despite the attractiveness of a smoothed and shifted output from each generator, this approach results in a significant decrease in the efficiency of the entire system, and eliminates the benefits of resource aggregation. For example, demand of any individual electricity consumer can be very irregular, with rapid unpredictable ramps. Distributed energy storage could be used to smooth these individual demands. However, this would result in storage devices being charged in one location, while simultaneously discharged in another, wasting both the cost of storage devices and the losses in the storage process. The aggregated net demand of many individual consumers is considerably smoother, and would require much less storage to "flatten" if desirable. Likewise, operational integration between energy storage and any individual or groups of VG would be nonoptimal and likely result in simultaneous charging and discharging. By aggregating the entire net load of a system, including all loads and VG supply, storage or other flexibility options can be deployed at the lowest cost and greatest efficiency.[68] This is especially the case when spatial diversity substantially reduces variability over multiple timescales.

There are some exceptions when there are benefits of operationally combining VG and energy storage, typically through co-location and sharing of certain high-cost components. The best example is integrating thermal storage into a concentrating solar power (CSP) plant; another example is sharing power electronics in a distributed PV/battery system.

There are several other applications where co-location of VG and storage may make sense. Wind plants placed in areas of weak transmission can potentially introduce power quality and stability issues, and storage can be a mitigating technology; however, improved power electronics in modern wind turbines may be a lower-cost alternative. Finally, combining wind and energy storage has been proposed as an alternative (or supplement) to developing new transmission capacity. Increased deployment of wind energy will require substantial new transmission, and storage co-located with remote wind resources can help decrease the need for new transmission.[69] This has been proposed to relieve congestion in the ERCOT grid, for example, where the state's best wind resources are located largely in the sparsely populated western part of the state, and transmission capacity is limited (Desai et al. 2003). Despite these potential applications, the majority of storage deployed in the grid will likely be a shared resource, which will benefit the entire system and not just a single generator or load (Smith et al. 2007). Just as loads are balanced in aggregate, the net load in the future grid – after all VG sources are included – will be balanced by a mix of conventional generation, plus flexibility options that include energy storage.

Table 5.1. Dedicated Renewable Applications of Energy Storage and Their Whole-Grid Counterpart

RE Specific Application	"Whole Grid" Application
Transmission Curtailment	Transmission Deferral
Time Shifting	Load Leveling/Arbitrage
Forecast Hedging	Forecast Error
Frequency Support	Frequency Regulation
Fluctuation Suppression	Transient Stability

5.3. Energy Storage Technologies and Applications

This section provides a brief overview of commercially available energy storage technologies. It is not intended to be a comprehensive discussion, because there are a large number of sources available that discuss the technical performance, current applications, vendors, and costs in detail. A number of more comprehensive reviews of energy storage technologies are provided in the Bibliography.

The choice of an energy storage device depends on its application in either the current grid or in the renewables/VG-driven grid; these applications are largely determined by the length of discharge. Energy storage applications are often divided into three categories, based on the length of discharge. Table 5.2 indicates the three regimes of energy storage applications commonly discussed.

The first two categories of energy storage applications in Table 5.2 correspond to a range of ramping and ancillary services, but do not typically require continuous discharge for extended periods of time. In the case of renewables-driven applications, this could require discharge times of up to about an hour to allow fast-start thermal generators to come online in response to forecast errors. (Bridging power typically refers to the ability of a storage device to "bridge" the gap from one energy source to another.)[70] The third category (energy management) corresponds to energy flexibility, or the ability to shift bulk energy over periods of several hours or more.

The references in the Bibliography provide a number of assessments and charts with estimates of technical performance and costs. Figure 5.2 provides one example of the range of technologies available for these three classes of services, and shows that many technologies can provide services across various timescales.

Table 5.2. Three Classes of Energy Storage

Common Name	Example Applications	Discharge Time Required
Power Quality	Transient Stability, Frequency Regulation	Seconds to Minutes
Bridging Power	Contingency Reserves, Ramping	Minutes to ~1 hour
Energy Management	Load Leveling, Firm Capacity, T&D Deferral	Hours

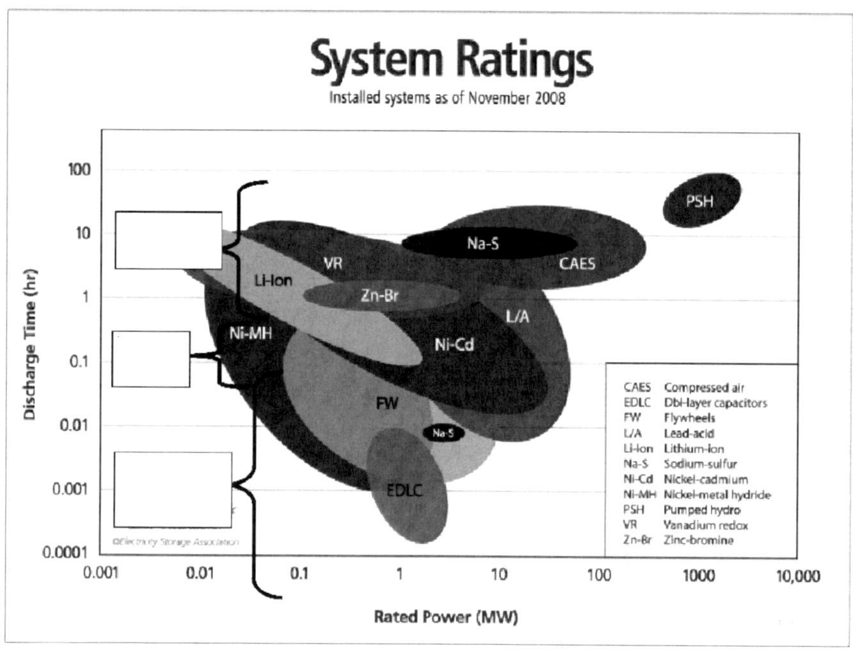

Figure 5.2. Energy storage applications and technologies[71]

It should be noted that this chart does not include thermal energy storage, which would cover a power range of a few kilowatts (kW) for thermal energy storage in buildings to more than 100 MW in CSP plants, with a discharge time of minutes to several hours.

When considering the technical performance or costs of energy storage, there are a number of caveats to consider. The first is the technical and commercial maturity of the storage technology. As of 2009, only four energy storage technologies (sodium-sulfur batteries, pumped hydro, CAES, and thermal storage)[72] have a total worldwide installed capacity that exceeds 100 MW.[73] This doesn't mean that there isn't market potential for any individual technology or storage, in general; but it makes it difficult to assess the state of any individual technology given its limited deployment to date. This also leads to some uncertainty in two of the primary performance indicators for energy storage devices: efficiency and cost. Both of these values are often imprecisely reported, which makes it difficult to perform an accurate assessment of the potential for individual technologies or compare different technologies. Some major caveats when considering electricity storage include:

Efficiency

1) The standard measure of an electricity storage device's efficiency in the grid is the AC to AC round-trip efficiency, or AC kWh_{out}/kWh_{in}.[74] However, this is not always the value reported, especially for devices that store DC energy such as batteries and capacitors. In some cases, the DC-DC round-trip efficiency may be reported, and additional losses in power conversion efficiencies must be considered if the device is to provide applications in the grid.
2) Reported round-trip efficiencies may not include "parasitic" loads. These include heating and cooling of batteries and power-conditioning equipment. These parasitic loads can vary considerably depending on use, climate, and the length of each storage cycle.
3) The round-trip efficiency of several technologies cannot be directly compared. Thermal storage provides some, but not all of the services of a "pure" electricity storage device; while compressed-air energy storage is a hybrid device that requires both electricity and natural gas. These factors limit the value of a direct comparison, as discussed in more detail later in this section.

Cost

1) As stated before, only a few storage technologies have been deployed at large scale (greater than 100 MW). Estimated prices for emerging technologies may be for a semi-custom product (and consequently very high) or projected costs based on mass production (and perhaps overly optimistic). Even with more mature technologies such as PHS and CAES, it has been some time since either has been built in the United States, so the cost of the next plant is somewhat uncertain.[75]
2) As with any generation technology, large variations in prices occur from year to year due to commodity prices and the global economy. Therefore, cost estimates of storage technologies from different years may reflect market conditions as opposed to real differences.
3) Storage technologies offer different classes of services and are comprised of an energy component and power component. The total cost of a storage device includes both components, with the limits of the target application. As a result, a direct comparison of a PHS device with a flywheel, for example, has limited value.

5.3.1. Storage Technologies for Power Quality Applications

Power quality applications require rapid response – often within less than a second – and include transient stability and frequency regulation. As with the other applications, the time scales of discharge may vary; but this class of services typically requires discharge times of up to about 10 minutes and nearly continuous cycling. Technologies for these applications include flywheels, capacitors, and superconducting magnetic energy storage (SMES).

Flywheels

Flywheels store energy in a rotating mass. Flywheels feature rapid response and high efficiency, making them well-suited for frequency regulation. Several flywheel installations have been planned or deployed to take advantage of high prices in frequency regulation markets (Lazarewicz 2009).

Capacitors

Capacitors[76] store electricity in an electric charge. Capacitors have among the fastest response time of any energy storage device, and are typically used in power quality applications such as providing transient voltage stability. However, their low energy capacity has restricted their use in longer time-duration applications. A major research goal is to increase their energy density and increase their usefulness in the grid (and potentially in vehicle applications.)

Superconducting Magnetic Energy Storage (SMES)

SMES stores energy in a magnetic field in a coil of superconducting material. SMES is similar to capacitors in its ability to respond extremely fast, but it is limited by the total energy capacity. This has also restricted SMES to "power" applications with extremely short discharge times. Several demonstration projects have been deployed.

5.3.2. Storage Technologies for Bridging Power

Bridging power applications include providing contingency reserves, load following, and additional reserves for issues such as forecast uncertainty and unit commitment errors. This set of applications generally requires rapid response (in seconds to minutes) and discharge times in the range of up to about an hour. Far less cycling is required than for power quality applications.

This application is generally associated with several battery technologies, which include lead-acid, nickel-cadmium, nickel-metal hydride, and (more recently) lithium-ion. Due to their rapid response, they can provide power quality services such as frequency regulation; but the continuous cycling requirement can limit battery life. Several demonstration projects have been built using these technologies to provide operating reserves.

5.3.3. Storage Technologies for Energy Management

Energy management applications include moving power over longer timescales, and generally require continuous discharge ratings of several hours or more. Technologies for these applications include several battery types, pumped hydro, compressed air, and thermal energy storage.

High-Energy Batteries

For many batteries, there is considerable overlap between energy management and the shorter-term applications discussed previously. Furthermore, batteries can generally provide rapid response, which means that batteries "designed" for energy management can potentially provide services over all the applications and timescales discussed.

Several battery technologies have been demonstrated or deployed for energy management applications. In addition to the chemistries discussed previously, the commercially available batteries targeted to energy management include two general types: high- temperature batteries and liquid electrolyte flow batteries

The most mature high-temperature battery as of 2009 is the sodium-sulfur battery, which has worldwide installations that exceed 270 MW (Rastler 2008). Alternative high- temperature chemistries have been proposed and are in various stages of development and commercialization. One example is the sodium-nickel chloride (ZEBRA) battery.

The second class of high-energy batteries is the liquid electrolyte "flow" battery. This battery uses a liquid electrolyte that flows across a membrane. The advantage of this technology is that the power component and energy component can be sized independently. As of 2009, there has been limited deployment of two types of flow batteries – vanadium redox and zinc-bromine. Other combinations such as polysulfidebromine have been pursed, and new chemistries are under development.

In the United States, a primary application of energy management batteries has been T&D deferral; however, demonstration projects have been deployed for multiple applications (Nourai 2007, EPRI 2003).

Pumped Hydro Storage (PHS)

Pumped hydro is the only energy storage technology deployed on a gigawatt scale in the United States and worldwide. In the United States, about 20 GW is deployed at 39 sites, and installations range in capacity from less than 50 MW to 2,100 MW.[77] Many of the sites store 10 hours or more, making the technology useful for load leveling. PHS is also used for ancillary services. PHS uses conventional pumps and turbines and requires a significant amount of land and water for the upper and lower reservoirs. PHS plants can achieve round-trip efficiencies that exceed 75% and may have capacities that exceed 20 hours of discharge capacity. Environmental regulations may limit large-scale aboveground PHS development. However, given the high round-trip efficiencies, proven technology, and low cost compared to most alternatives, conventional PHS is still being pursued in a number of locations.[78] Alternative lower-impact configurations have been studied, including using a natural or mined underground formation for the lower reservoir, but this configuration has yet to be commercialized.

Compressed Air Energy Storage (CAES)

CAES technology is based on conventional gas turbine technology and uses the elastic potential energy of compressed air. Energy is stored by compressing air in an airtight underground storage cavern. To extract the stored energy, compressed air is drawn from the storage vessel, heated, and then expanded through a high-pressure turbine that captures some of the energy in the compressed air. The air is then mixed with fuel and combusted, with the exhaust expanded through a low-pressure gas turbine. The turbines are connected to an electrical generator.

CAES is considered a hybrid generation/storage system because it requires combustion in the gas turbine. The performance of a CAES plant is based on its energy ratio (energy in/energy out) and its fuel use (typically expressed as heat rate in BTU/kWh). CAES performance is estimated at an energy ratio of 0.6-0.8 and a heat rate of 4,000-4,300 BTU/kWh (Succar and Williams 2008). Because CAES uses both electricity and natural gas, a single-point definition of the round-trip efficiency of a CAES device does not represent an economic figure of merit.

The primary disadvantages of CAES are the need for an underground cavern and its reliance on fossil fuels. Alternative configurations for CAES have been proposed using manufactured above-ground vessels, new turbine designs to reduce fossil fuel use, or designs that re-use the heat of compression and avoid fuel use altogether.

Thermal Energy Storage

Thermal energy storage is sometimes ignored as an electricity storage technology because it typically is not used to store and then discharge electricity directly. However, in some applications, thermal storage can be functionally equivalent to electricity storage. One example is storing thermal energy from the sun that is later converted into electricity in a conventional thermal generator. Another example is converting electricity into a form of thermal energy that later substitutes for electricity use such as electric cooling or heating.

The first example (storing thermal energy that is later converted into electricity) can be used with many types of thermal generators but is most often associated with concentrating solar power. In this application, thermal energy from the solar field is stored in molten salt or another medium. This energy can be recovered later and used to generate electricity, which turns this technology into a dispatchable source of energy.

Care must be used when discussing the efficiency of thermal energy storage. One of the major issues with electricity storage is efficiency losses. Electricity is a high "quality" source of energy, and transforming electricity into a stored medium and back incurs considerable losses. Thermal energy is a much lower quality of energy, but can be stored with much higher efficiency. In a CSP plant, thermal energy is stored before conversion to electricity. As a result, the round-trip efficiency of CSP thermal storage may be close to 100%, much higher than any electricity storage technology. However, CSP thermal storage can only store thermal energy produced from the solar field, as opposed to other storage technologies that can store electricity produced from any source.

Likewise, end-use energy storage can have extremely high round-trip efficiencies. Demand for electric-power cooling can be shifted by storing cold energy in the form of chilled water or ice during off-peak times and releasing that cold energy during times of peak demand. This effectively stores electricity with high round-trip efficiency.[79]

End-use hot storage can also be used in both space heating and water heating applications. (Controllable water heating somewhat blurs the line between energy storage and demand response.) As with other forms of thermal storage, the effective round-trip efficiency of end-use hot storage is much higher than "pure" electricity storage devices but is limited by daily and seasonal heating demands.

5.4. Electric Vehicles and the Role of Vehicle to Grid

Electric vehicles (EVs – used here to represent both "pure" electric vehicles or plug-in hybrid electric vehicles) are a potential source of flexibility for VG applications. The charging of EVs can potentially be controlled, and provide a source of dispatchable demand and demand response. Controlled charging can be timed to periods of greatest VG output, while charging rates can be controlled to provide contingency reserves or frequency regulation reserves. Vehicle to grid (V2G) (where EVs can partially discharge stored energy to the grid) may provide additional value by acting as a distributed source of storage. EVs could potentially provide all three grid services discussed previously. Most proposals for both controlled charging and V2G focus on short-term response services such as frequency regulation and contingency. Their ability to provide energy services is more limited by both the storage capacity of the battery, as well as the high cost of battery cycling. This could restrict their ability to provide time-shifting (energy arbitrage) beyond their ability to perform controlled charging. [80] The role of V2G is an active area of research, and because electrified vehicles in any form have yet to achieve significant market penetration, it is difficult to assess their potential as a source of grid flexibility. However, analysis has demonstrated potential system benefits of both controlled charging and V2G (Denholm and Short 2006). The role of EVs as an enabling technology requires additional analysis of their unique temporal characteristics of availability, unknown battery costs and lifetimes, and the availability of smart charging stations to maximize their usefulness while parked.

6. CONCLUSIONS

The increasing role of variable renewable sources (such as wind and solar) in the grid has prompted concerns about grid reliability and raised the question of how much these resources can contribute before enabling technologies such as energy storage are *needed*. Fundamentally, this question is overly simplistic. In reality, the question is an economic issue: It involves the integration costs of variable generation and the amount of various storage or other enabling technologies that are economically viable in a future with high penetrations of VG. To date, integration studies of wind to about 20% on an energy basis have found that the grid can accommodate a substantial increase in VG without the need for energy storage, but it will require changes in operational practices, such as sharing of generation resources and loads over larger areas. Beyond this level, the impacts and costs are less clear, but 30% or more appears feasible with the introduction of "low-cost" flexibility options such as greater use of demand response. However, these studies have not necessarily focused on storage and generally do not attempt to determine the optimal system (including the amount of storage) that provides the lowest cost of energy.

There are technical and economic limits to how much of a system's energy can be provided by VG without enabling technologies based on at least two factors: coincidence of VG supply and demand and the ability to reduce output from conventional generators. At extremely high penetration of VG, these factors may cause excessive (and costly) curtailment, which will require methods to increase the useful contribution of VG However, the concern regarding how much VG can be used before storage is the most economic option for further integration currently has no simple answer, primarily because the availability and cost of grid flexibility options are not well understood and vary by region.

It is clear that high penetration of variable generation increases the need for all flexibility options including storage, and it also creates market opportunities for these technologies. Historically, storage has been difficult to sell into the market, not only due to high costs, but also because of the array of services it provides and the challenges it has in quantifying the value of these services – particularly the operational benefits such as ancillary services. The challenge of simulating energy storage in the grid, estimating its total value, and actually recovering those value streams continues to be a major barrier. VG complicates this issue because variability adds additional analysis challenges. The ability to simulate the cost impacts of VG and benefits of storage is still limited by the methods and data sets available. It is understood that VG increases the need for flexible generation and operating reserves, which can be met by energy storage. However, the value of energy storage is best captured when selling to the entire grid, instead of any single source. Evaluating the role of storage with VG sources requires continued analysis, improved data, and new techniques to evaluate the operation of a more dynamic and intelligent grid of the future.

Bibliography

The following are general overviews and reviews of energy storage technologies that provide additional information.

American Physical Society (APS). (2007). Challenges of Electricity Storage Technologies. Accessed December 2009 at *http://www.aps.org/policy/ reports/popareports/upload/Energy-2007-Report-ElectricityStorageReport.pdf.*

Chi-Jen, Y. & Williams, E. (2009). "Energy Storage for Low-Carbon Electricity." Duke University Climate Change Policy Partnership. http://www.nicholas.duke.edu/ccpp/ccpp pdfs/energy.storage.pdf.

Electric Power Research Institute/Department of Energy (EPRI/DOE). (2003). *Handbook of Energy Storage for Transmission and Distribution Applications.* Palo Alto, CA: 2003. 1001834.

Electricity Advisory Committee. (2008). *Bottling Electricity: Storage as a Strategic Tool for Managing Variability and Capacity Concerns in the Modern Grid.* Accessed December 2009 at *http://www.oe.energy. gov/DocumentsandMedia/final-energy-* storage 12-1 6-08.pdf.

Electricity Storage Association. (2009). Accessed December 2009 at http://www.electricitystorage.org/site/home/.

Hall, P. J. & Bain, E. J. (2008). "Energy-storage technologies and electricity generation." *Energy Policy, 36*, 4352–4355.

Ibrahima, H., Ilincaa, A. & Perron, J. (2008). "Energy storage systems—characteristics and comparisons." *Renewable and Sustainable Energy Reviews, 12*, 1221–1250.

Pew Center on Global Climate Change. (2009). "Electric Energy Storage." Accessed December 2009 at *http://www.pewclimate.org/ technology/ factsheet/EnergyStorage.*

Rastler, D. (2008). "New Demand for Energy Storage," September/October 2008. Accessed December 2009 at *http://www.eei.org/magazine/ EEI%20Electric%20Perspectives%20Article%20Listing/20* 08-09-0 1 - EnergyStorage.pdf.

References

Ackermann, T., Ancell, G., Borup, L. D., Eriksen, P. B., Ernst, B., Groome, F.,

Lange, M., Mohrlen, C., Orths, A.G., O'Sullivan, J. & de la Torre, M. (2009). "Where the wind blows." *IEEE Power and Energy Magazine*, Vol. 7, No. 6. (30 October 2009), pp. 65-75.

Adamson, D. M. (2009). "Realizing New Pumped-Storage Potential Through Effective Policies," *Hydro Review*, April 2009, pages 28-30.

American Society of Civil Engineers (ASCE). (1993). Task Committee on Pumped Storage of the Hydropower Committee of the Energy Division of the American Society of Civil Engineers. "*Compendium of Pumped Storage Plants in the United States.*" American Society of Civil Engineers, New York.

American Wind Energy Association. (2009). "Annual Wind Industry Report Year Ending 2008." Accessed December 2009 at *http://www. awea.o rg/publications/reports/AWEAAnnual-Wind-Report-2009.pdf*

Boyd, D. W., Buckley, O. E. & Clark, C. E. (1983). "Assessment of Market Potential of Compressed-Air Energy-Storage Systems." *Journal of Energy,* 1983, *7*, 549-556.

Butler, P. C., Iannucci, J. & Eyer, J. (2003). *Innovative Business Cases For Energy Storage In a Restructured Electricity Marketplace,* SAND2003 - 0362 February 2003.

California Independent System Operator (CAISO). (2007). "Integration of Renewable Resources." November 2007. California Independent System Operator.

Corbus, D., Milligan, M., Ela, E., Schuerger, M. & Zavadil, B. (2009). Eastern Wind Integration and Transmission Study-Preliminary Findings: Preprint. 9 pp., NREL Report No. CP-550-46505.

DeCesaro, J., Porter, K. & Milligan, M. (2009). "Wind Energy and Power System Operations: A Review of Wind Integration Studies to Date." *The Electricity Journal, Volume 22*, Issue 10, Pages 34-43 (December 2009).

Denholm, P. & Sioshansi, R. (2009). "The Value of Compressed Air Energy Storage with Wind in Transmission-Constrained Electric Power Systems," *Energy Policy, 37*, 3149-3 158.

Denholm, P. & Margolis, R. M. (2008). "Supply Curves for Rooftop Solar PV-Generated Electricity for the United States," NREL/TP-670-44073.

Denholm, P. & Margolis, R. M. (2007a). "Evaluating the Limits of Solar Photovoltaics (PV) in Traditional Electric Power Systems" *Energy Policy, 35*, 2852-2861.

Denholm, P. & Margolis, R. M. (2007b). "Evaluating the Limits of Solar Photovoltaics (PV) in Electric Power Systems Utilizing Energy Storage and Other Enabling Technologies." *Energy Policy, 35*, 4424-443 3.

Denholm, P. & Letendre, S. E. (2007). "Grid Services from Plug-in Hybrid Electric Vehicles: A Key to Economic Viability?" *Electrical Energy Storage–Applications and Technology Conference*. September, *25*, 2007.

Denholm, P. & Short, W. (2006). "An Evaluation of Utility System Impacts and Benefits of Optimally Dispatched Plug-In Hybrid Electric Vehicles." NREL/TP-620-40293.

Department of Energy (DOE). (1977). DOE Interagency Coordination Meeting on Energy Storage, CONF 7709116.

Desai, N., Nelson, S., Garza, S., Pemberton, D. J., Lewis, D., Reid, W., Lacasse, S., Spencer, R., Manning, L. M. & Wilson, R. (2003). "Study Of Electric Transmission In Conjunction With Energy Storage Technology," Lower Colorado River Authority, Texas State Energy Conservation Office, August 21, 2003.

Doherty, R., Mullane, A., Nolan, G., Burke, D. J., Bryson, A. & O'Malley, M. (forthcoming) "An Assessment of the Impact of Wind Generation on System Frequency Control." *IEEE Transactions on Power Systems,* forthcoming.

Energy Information Administration (EIA). (2009a). *Electric Power Annual,* 2009 (data for 2008). Accessed December 2009 at *http://www.e ia.doe.gov/cneaf/electricity* sprdshts.html

EIA. (2009b). "Repeal of the Powerplant and Industrial Fuel Use Act (1987)" Accessed December 2009 at http://www.eia.doe.gov/oil analysis publications/ngmajorleg/repeal.html.

Ela, E. & Kirby, B. (2008). "ERCOT Event on February 26, 2008: Lessons Learned." National Renewable Energy Laboratory. NREL/TP-500-43373, July 2008.

EnerNex Corporation. (2009). "Solar Integration Study for Public Service Company of Colorado." February 2009 for Xcel Energy. Accessed December 2009 at *http://www.xcelenergy.com/SiteCollectionDocument s/docs/PSCo_SolarIntegration_0209* 09.pdf

Electric Power Research Institute/Department of Energy (EPRI/DOE). (2004). *Energy Storage for Grid Connected Wind Generation Applications*, EPRI-DOE Handbook Supplement 1008703, December 2004.

EPRI. (2003). Handbook of Energy Storage for Transmission and Distribution Applications. *Palo Alto*, CA: 2003. 1001834.

EPRI. (1976). "Assessment of energy storage systems suitable for use by electric utilities." EPRI-EM-264, July 1976.

Electricity Storage Association (ESA). (2009). "Technology Comparisons." Accessed December 2009 at *http://www.electricitystorage.org/site/*

technologies/

European Wind Energy Association (EWEA). (2009). "Global Installed Wind Power Capacity." Accessed December 2009 at *http://www.ewea.or g/fileadmin/ewea_documents/documents/press_releases/2009/GWEC* Press Release - tables and statistics 2008.pdf

Eyer, J. M., Iannucci, J. J. & Corey, G. P. (2004). "*Energy Storage Benefits and Market Analysis Handbook*." A Study for the DOE Energy Storage Systems Program. SAND2004-6177, December 2004.

Figueiredo, F. C., Flynn, P. C. & Cabral, E. A. (2006). "The economics of energy storage in 14 deregulated power markets." *Energy Studies Review, 14*, 131–152.

General Electric (GE) Energy. (2008). "Analysis of Wind Generation Impact on ERCOT Ancillary Services Requirements." Prepared for the Electric Reliability Council of Texas (ERCOT), March 2008. Accessed December 2009 at http://www.uwig.org/AttchAERCOT A-S Study Exec Sum.pdf

Hurlbut, D. J. (2009). *Colorado's Prospects for Interstate Commerce in Renewable Power.* Prepared for the Colorado Governor's Energy Office, Renewable Energy Development Infrastructure (REDI). NREL Report No. TP-6A2-47179.

Independent System Operators and Regional Transmission Organizations Council (IRC). (2009). "2009 State of the Markets Report," prepared by the ISO/RTO Council. Accessed December 2009 at *http://www.isorto.or g/atf/cf/%7B5B4E85C6-7EAC-40A0-8DC3-* 00382951 8EBD%7D/2 009%20IRC%20State%20of %20Markets% 20Report.pdf

Jabbour, S. J. & Wells, W. M. (1992). "Optimal Dispatching of Storage Plants with Dynamics," in *Proceedings: Second International Conference on Compressed Air Energy Storage.* December 1992. EPRI TR-101770.

Kirby, B. (2004). "*Frequency Regulation Basics and Trends*," Oak Ridge National Laboratory, December 2004, ORNL/TM 2004/291.

Lamont, A. D. (2008). "Assessing the long-term system value of intermittent electric generation technologies." *Energy Economics, 30*, No. 3 (May), 1208-1231.

Lazarewicz, M. (2009). "*Commercial Flywheel Based Frequency Regulation Status*," Electricity Storage Association annual meeting 2009. Washington, D.C., May 21-22, 2009.

Lefton, S. A. & Besuner, P. (2006). "The Cost of Cycling Coal Fired Power Plants," *Coal Power Magazine*, Winter.

Lew, D., Milligan, M; Jordan, G., Freeman, L., Miller, N., Clark, K. & Piwko, R. (2009). "How do Wind and Solar Power Affect Grid Operations: The

Western Wind and Solar Integration Study." NREL/CP-550-46517.

Makarov, Y. V., Ma, J., Lu, S. & Nguyen, T. B. (2008). "*Assessing the Value of Regulation Resources Based on Their Time Response Characteristics*." June 2008, PNNL–17632.

Milligan, M., Lew, D., Corbus, D., Piwko, R., Miller, N., Clark, K., Jordan, G., Freeman, L., Zavadil, B. & Schuerger, M. (2009a). *Large-Scale Wind Integration Studies in the United States: Preliminary Results*. NREL/CP-550-46527, September 2009.

Milligan, M., Kirby, B., Gramlich, R. & Goggin, M. (2009b). "The Impact of Electric Industry Structure on High Wind Penetration Potential." NREL/TP-550-46273, July 2009.

Milligan, M. & Kirby, B. (2009). *Calculating Wind Integration Costs: Separating Wind Energy Value from Integration Cost Impacts*. NREL TP-550-46275.

Nickell, B. M. (2008). "Wind Dispatchability and Storage Interconnected Grid Perspective." Presentation accessed December, at *http://www.austinenergy.com/About%20Us/Newsroom/Reports/taskForce/tfWindDispatc* hability.pdf.

North American Electric Reliability Corporation (NERC). (2009). "Accommodating High Levels of Variable Generation," April 2009. Accessed December 2009 at http://www.nerc.com/docs/pc/ivgtf/IVGTF Outline Report 04070 8.pdf.

NERC. (2008). "NERC Glossary of Terms Used in Reliability Standards," February 12, 2008. Accessed December 2009 at *www.nerc.com/files/ Glossary* 12Feb08.pdf.

Northwest Power and Conservation Council. (2009). "Draft Northwest Sixth Power Plan," September. *http://www.nwcouncil.org/energy/powerplan/6/DraftSixthPowerPlan* 0903 09.pdf

Nourai, A. (2007). "Installation of the First Distributed Energy Storage System (DESS) at American Electric Power (AEP)," June 2007, SAND2007-3580.

Nourai, A., Kogan, V. I. & Schafer, C. M. (2008). "Load Leveling Reduces T&D Line Losses," *IEEE Transactions on Power Delivery*, October 2008, Volume: *23*, Issue: 4, 2168-2173.

Palmintier, B., Hansen, L. & Levine, J. (2008). "Spatial and Temporal Interactions of Solar and Wind Resources in the Next Generation Utility," Solar 2008, San Diego, CA.

Peterson, S. B., Whitacre, J. F. & Apt, J. (2010). "The Economics of Using PHEV Battery Packs for Grid Storage." *Journal of Power Sources*, *195*,

2377-2384.

PJM Interconnections (PJM). (2009a). "PJM Manual 12: Balancing Operations," Revision: 20, October 5, 2009. *http://www.pjm.com/ markets-and-operations/ancillaryservices/~/media* 2.ashx

PJM. (2009b). "PJM Manual 11: Scheduling Operations," Revision: 43. September 24, 2009. *http://www.pjm.com/markets-and-operations/ancill aryservices/~/media* 1 .ashx

Rastler, D. (2008). "New Demand for Energy Storage," *Electric Perspectives,* September/October 2008. Accessed December 2009 at *http://www.eei.o rg/magazine/EEI%20Electric%20Perspectives%20Article%20Listing/20* 08-09-0 1 -EnergyStorage.pdf

Rebours Y. & Kirschen, D. S. (2005). "*A survey of definitions and specifications of reserve services*," University of Manchester.

Sherwood, L. (2009). "U.S. Solar Market Trends 2008." Interstate Renewable Energy Council, published July 2009. Accessed December 2009 at http://www.irecusa.org/fileadmin/user_upload/ NationalOutreach Docs/So larTrendsReport s/IREC Solar Market Trends Report 200 8.pdf .

Short, W. & Denholm, P. (2006). "*A Preliminary Assessment of Plug-In Hybrid Electric Vehicles on Wind Energy Markets*," NREL/TP-620-39729.

Silverman, J. (ed). (1980). "*Energy Storage: a vital element in mankind's quest for survival and progress*." Transactions of the First International Assembly held at Dubrovnik, Yugoslavia, 27 May-1 June, 1979. Pergamon Press.

Sioshansi, R., Denholm, P., Jenkin, T. & Weiss, J. (2009) "Estimating the Value of Electricity Storage in PJM: Arbitrage and Some Welfare Effects," *Energy Economics*., *31*, 269-277.

Smith, J. C. & Parsons, B. (2007). "What Does 20% Look Like?" IEEE Power & Energy, Nov/Dec 22-33.

Smith, J. C., Milligan, M. R., DeMeo, E. A. & Parsons, B. (2007). "Utility Wind Integration and Operating Impact State of the Art," *IEEE Transactions On Power Systems*, Vol. *22*, No 3, August 2007.

Strauss, P. L. (1991). "Pumped storage, the environment and mitigation." *Waterpower, 91*, New York

Succar, S. & Williams, R. H. (2008). "Compressed Air Energy Storage: Theory, Resources, And Applications For Wind Power," published for the Princeton Environmental Institute, April.

Troy, N. E., Denny, E., M. O'Malley, M. (forthcoming) "Base-load Cycling on a System with Significant Wind Penetration" *IEEE Transactions on*

Power Systems, forthcoming.

Walawalkar, R., Apt, J. & Mancini, R. (2007). "Economics of electric energy storage for energy arbitrage and regulation in New York," *Energy Policy, 35*, 2558–2568.

Zavadil, R. (2006). "Wind Integration Study for Public Service Company of Colorado," Enernex Corporation for Xcel Energy, May 2006.

End Notes

[1] The use of the term "intermittent" has been questioned by the wind energy community as being technically inaccurate. Intermittent implies a short-term "on-off" cycle while the output of wind experiences maximum variations more typically on the order of 10% per hour. Solar PV is perhaps somewhat more "intermittent" because it follows a daily on-off cycle. The description "variable" or "variable and uncertain" has been proposed as a more technically accurate description of the output of a wind power plant (Smith and Parsons 2007).

[2] Most of Texas (about 85% of the population) is within the ERCOT grid, which is largely independent of the two larger U.S. grids.

[3] Operating reserves are primarily capacity services (the ability to provide energy on demand) as opposed to actual energy services.

[4] The nomenclature around various ancillary services (especially spinning reserves) varies significantly. While the NERC glossary indicates that spinning reserve applies to both contingency and frequency regulation, the term spinning reserve often is used to refer to only contingency reserves. For additional discussion of nomenclature around contingency and spinning reserves, see Rebours and Kirschen 2005.

[5] This "opportunity cost" associated with uneconomic dispatch is the dominant source of reserve costs (Kirby 2004).

[6] PHS stores energy by pumping water from a lower reservoir to an upper reservoir and releasing that stored water through a conventional hydroelectric generator. Additional information about PHS is provided in Section 5.

[7] Concerns about the availability of oil and other peaking fuels in this period was so great that an international conference (including the U.S. National Academy of Sciences) on the subject in 1979 described energy storage as "a vital element in mankind's quest for survival and progress" (Silverman 1980).

[8] See, for example, EPRI 1976. Here, the proposed method for comparing energy storage to conventional alternatives is based solely on the value of energy and firm capacity value without any actual quantification of operational benefits.

[9] This would be a typical assumption for a pumped hydro plant built during the 1970s and 1980s (EPRI 1976).

[10] This figure is intended to represent general trends as opposed to absolute costs. In this figure, fuel prices and generation characteristics are derived from various data sets from the Energy Information Administration.

[11] PHS also takes longer to build (increasing the risk for investors), requires additional permits, and is typically located farther from load centers, which requires more transmission than gas-fired generators. PHS may also face greater environmental opposition (Strauss 1991).

[12] This problem has been noted many times. For example, "traditionally, when electric utilities evaluate generating additions to their facilities, the evaluation process considers the contribution of each alternative to both capacity and energy requirements. However, the evaluation process often neglects or inaccurately measure potential costs and benefits not

directly related to capacity and energy. Operating considerations that reflect the ability (or inability) of a generation resource to respond to the electric system's dynamic operating needs usually fall into this category." (Jabbour and Wells 1992).

[13] Private investors tend to favor lower capital cost investments with faster construction times (i.e., combustion turbines and combined-cycle plants), even if they have higher operating costs, because this reduces perceived economic risk.

[14] To place these values in perspective, between 1993 and 2008, more than 320 GW of conventional capacity was constructed in the United States. With the exception of the completion of previously started PHS facilities and a few demonstration projects, no significant storage capacity was added. The total U.S. utility storage capacity of about 20 GW in 2008 is less than 2% of the total installed generation capacity (EIA 2009a).

[15] This study analyzed devices up to 40 hours, and found rapidly diminishing returns for devices with storage capacity greater than about 10 hours. The majority of arbitrage benefits are within a day, as opposed to over larger time periods. In addition, while short-term price variation is highly predictable, long-term variations are less predictable, which reduces the certainty of long-term arbitrage opportunities.

[16] This is the inverse of the process of calculating an annualized cost from a capital cost by multiplying by the capital charge rate.

[17] These values also assume that frequency regulation is an energy- and cost-neutral service.

[18] Capital charge rate (CCR) of 9.8% from EPRI 2003, CCR of 12% from Butler et al. 2003, CCR of 13.9% from Eyer et al. 2004.

[19] CAES is one possible exception, but requires analysis of both the electricity price and natural gas prices. See Section 5 and the references in the Bibliography for additional discussion of storage costs.

[20] PJM's capacity market (Reliability Pricing Model) data derived from *http://www.pjm.com/markets-andoperations/rpm.aspx.*

[21] More recent data shows a greater range of value for contingency reserves. For example, in 2008, the average value for spinning reserves in NYISO was $10.1 in the east and $6.2 in the west, corresponding to as little as $54/kW-yr. In the same year, the average price in ERCOT was $27.1, corresponding to $237/kW-yr. In 2009, these values fell substantially to $36/kW-yr in NYISO west, and $87/kW-yr in ERCOT.

[22] Frequency regulation theoretically is a net zero energy service over relatively short time scales, meaning the energy capacity of the device can be much smaller than those providing operating reserves and energy arbitrage. Several markets in the United States have changed or have proposed to change their treatment of regulation to accommodate energy-limited storage technologies. Furthermore, it has been suggested that fast-responding storage devices could receive a greater value per unit of capacity actually bid, because they could actually reduce the amount of reserves needed. For example, "faster responsive resources can help to reduce California ISO's regulation procurement by up to 40% (on average)" and "California ISO may consider creating better market opportunities and incentives for fast responsive resources." (Makarov et al. 2008).

[23] Large PHS units or other black-start generators must themselves be "black-started" and may use batteries or small generators for this purpose. Many transmission substations also use batteries partly to maintain reliability during power failures.

[24] Contingency reserves may be provided by both spinning and non-spinning units, depending on the market. The requirements for non-spinning reserves are the same except the resource does not need to "begin responding immediately." Full response is still within 10 minutes.

[25] For example, in the PJM regional transmission organization (RTO) in 2008 (covering about 50 million people), synchronized reserves were called a total of 40 times with an average duration of 10 minutes. See http://www.pjm.com/markets-and-operations/ancillary-services.aspx

[26] The black-start performance standard in PJM is 90 minutes to start, with an ability to run for 16 hours (PJM 2009a).

[27] In some locations such as Texas, demand response typically provides half of the contingency reserve requirements. Other regions also use (or are evaluating) load to provide regulation.

[28] From this point on, variable renewable generators will be referred to as variable generation (VG) following NERC 2009.

[29] This figure uses ERCOT load data from 2005 along with 15 GW of spatially diverse simulated wind data from the same year. See Section 4.2 for more details about the data used.

[30] A reduction in demand from VG or load will reduce the total marginal cost of generation, which includes fuel, emissions costs, and variable O&M. This ignores any additional cost impacts of variability on the remaining generation fleet, which is discussed in the next section.

[31] There are a number of other benefits provided by renewable energy sources such as reduced volatility of fuel prices. This work is not intended to be an analysis of the total benefits of renewables or VG.

[32] As discussed later, the impact of short-term wind variability is often overstated, especially considering the benefits of spatial diversity. The impact on minute-to-minute regulation requirements is mitigated by aggregating large amounts of wind because individual wind plant variability is uncorrelated in the regulation time frame. Furthermore, the limited ability to forecast wind and use of persistence forecasts may be a major factor in increased short-term variability of the net load. Finally, newer wind turbines meeting "low-voltage ride-through" standards can add short-duration stability, and have the capability to provide frequency regulation to the grid.

[33] This is actually the combination of the uncertainty in load and the uncertainty in wind. As VG penetration increases, it begins to dominate the net load uncertainty.

[34] These tools have several names such as "production cost" or "security-constrained unit commitment and economic dispatch" models.

[35] This actually oversimplifies the situation. Some research indicates that these costs are not entirely integration costs but a modeling artifact of how the "base case" in these studies is actually simulated (Milligan and Kirby 2009).

[36] "Combining geographically diverse wind and solar resources into a single portfolio tends to reduce hourly and sub-hourly variations in real-time output. This would result in a more consistent level of output over a longer time frame, which could reduce the cost of wind integration" (Hurlbut 2009). For additional analysis of spatial diversity, see also Palmintier et al. 2008.

[37] In 2008, more than 8.5 GW of wind was installed in the United States, reaching a total capacity of about 25.3 GW by the end of the year (AWEA 2009). In the same year, 0.3 GW of solar PV was installed, reaching a total capacity of about 1 GW (Sherwood 2009).

[38] The impacts of CSP are largely unquantified as well. However, it is expected that the impact of CSP on short time scales will be significantly less than PV, because CSP has significant "thermal inertia" in the system that will minimize high-frequency ramping events. Furthermore, CSP has the potential advantage of utilizing high-efficiency thermal storage discussed in Section 5.

[39] It should be noted that these increased costs are associated with the "residual" part of the system that provides the load not met by VG. This is often a source of confusion and is sometimes interpreted as an increase in fuel use and emissions of the total system – implying that the additional reserve requirements of VG somehow actually increase fuel use and emissions of the system, or that VG has a net negative impact on emissions. This is not the case, and integration studies have universally concluded that any increase in fuel use associated with reserves for VG is much smaller than overall avoided fuel and emissions from displaced conventional generation.

[40] Customers with interruptible loads that can meet certain performance requirements may be qualified to provide operating reserves under the Load Acting as a Resource (LaaR) program. In eligible ancillary services (AS) markets, the value of the LaaR load reduction is

equal to that of an increase in generation by a generating plant. See *http://www.ercot.com/services/programs/load/laar/*

41 This issue has important implications for the use of storage or any other device used to mitigate uncertainty. Energy storage, like any other generator, needs to be scheduled – a storage device used for load leveling may not be able to simultaneously provide hedging against under-forecasted wind, because it may already be discharging.

42 A number of utilities have expressed the opinion that these costs are not well-captured in previous wind integration studies. This issue is discussed later in this chapter.

43 This also explains why no significant new storage has been developed in the United States or Europe despite the 25 GW and 65 GW of wind development, respectively, as of the end of 2008 (EWEA 2009).

44 The more recent U.S. studies of very high penetration (the Western Wind and Solar Integration Study and the Eastern Wind Integration Study) require power and energy exchanges over larger areas than typically occur in the existing system (Milligan et al. 2009a).

45 One additional challenge in a high-VG grid is the potential decrease in mechanical inertia that helps maintain system frequency. This concern is not well understood and could be mitigated by a variety of technologies including improved controls on wind generators, or other sources of real or virtual inertia that could include energy storage. See, for example, Doherty et al. (forthcoming).

46 This simulation uses historical load data from ERCOT from 2005 and simulated wind data for the same year provided by AWS Truewinds. While the wind and load data is from ERCOT, the mix of generators (flexible and inflexible) is hypothetical and used only to illustrate the impact of VG. For additional discussion of the wind data, see GE Energy 2008.

47 The alternative to curtailment is finding some alternative use for this energy through enabling techniques and technologies, which may include energy storage. These options are discussed in Section 5.

48 "During over-generation periods, when dispatchable generation plants are already operating at their minimum levels, the California ISO needs to have an ability to curtail wind generation on an as-needed basis." (CAISO 2007)

49 When curtailing output, wind or other VG can supply operating reserves. In some cases, the value of curtailed energy may actually exceed the value of energy, but the primary value of VG is displacing conventional generation. While Figure 4.3 curtails renewable generation, it not clear which plant "should" be curtailed. Ignoring operational constraints, from a strict economic sense, VG has a lower cost of energy and lower emissions; therefore, it should be curtailed last.

50 "Cycling operations, that include on/off startup/shutdown operations, on-load cycling, and high frequency MW changes for automatic generation control (AGC), can be very damaging to power generation equipment." However, these costs can be very difficult to quantify, especially isolating the additional costs associated with cycling above and beyond normal operations (Lefton et al. 2006).

51 Wind integration studies typically use proprietary software and data sets, and do not always state which costs are and are not included. However, in the Western Wind and Solar Integration Study (the highest penetration U.S. integration study as of 2009) the study states: "'Wear and tear' costs due to increased or harder cycling of units were not taken into account because these have not been adequately quantified." (Lew et al. 2009).

52 A more detailed analysis of the relationship between wind penetration and plant cycling is provided by Troy et al. (forthcoming).

53 During this year, the average price paid for coal in this region was about $31/ton. With a typical heat rate of about 10,500 BTU, this translates into a variable cost of $12-14/MWh. This cost can be observed in Figure 4.4 as the "floor" cluster of points at this level. Points below this level represent bids that are less than the cost of generation, but not necessarily uneconomic if the alternative is excessive cycling-induced maintenance or even a forced shutdown and very expensive restart of a coal generator.

[54] Minimum load points would be less of a constraint if power plants could be quickly shut down and started up at low costs. With the exception of certain peaking plants such as aeroderivative turbines and fast-starting reciprocating engines, most plants have minimum up-and-down times, and require several hours to restart (at considerable cost).

[55] This is noted in previous wind integration studies such as the ERCOT study (GE 2008), and the more recent Eastern and Western Interconnect studies. "As the penetration of wind and solar increase, the impact on base-load coal increases, becoming very challenging at the 30% penetration." (Milligan et al. 2009b).

[56] Historically, many estimates of the limits of wind penetration have used data from a single or very small set of wind power plants, and often a small balancing area. Without spatial diversity of resource and load, this leads to both excessive ramp rates and excessive curtailment.

[57] The data set is the same as from Figures 4.2 and 4.3, with solar data derived from the National Solar Radiation Database. See Denholm and Margolis 2008 for additional details. The data was processed using the REFlex Model, which compares hourly electricity demand with VG supply and calculates curtailments as a function of system flexibility based on minimum load constraints. The model is described in more detail in Denholm and Margolis 2007a and 2007b.

[58] It should be noted that significant changes in the generation mix and corresponding changes in system flexibility could result from CO_2 emissions constraints. Increased cost of carbon could motivate greater use of natural gas generation, and reduce use of coal, both of which would tend to increase system flexibility, and allow greater economic use of VG. This, in turn, would decrease fuel use and the capacity factor of thermal generators, which would also tend to increase the use of gas-fired generation and decrease the use of coal as the economic optimal mix of generation. For additional discussion, see Lamont 2008.

[59] The large number of options available for increasing the penetration of RE is one of the reasons why the question of when storage becomes necessary is very difficult to answer. There is less of a technical limit than an economic one that depends on a large number of factors such as the cost of storage compared to a vast array of alternatives.

[60] The data and methods used for this analysis are the same as before, using the REFlex model to place otherwise curtailed energy into storage and using that stored energy at a later time (Denholm and Margolis 2007b).

[61] Alternatively, for a given amount of allowed curtailment, the contribution of VG increases with the addition of storage. In the case illustrated in Figure 4.11, a maximum VG curtailment of 10% allows VG to provide about 35% of the total electricity demand. Adding energy storage (with its ability to reduce the minimum load) increases the contribution of VG (for the same amount of allowable curtailment) to about 42% of total demand.

[62] Based on an original by Nickell 2008. While the figure hypothesizes an order of the supply curve, the actual costs and availability of the individual components are conjectural.

[63] "Larger markets and balancing areas that are a central feature of ISOs and RTOs can improve the physical conditions needed to integrate large amounts of wind energy. ISOs and RTOs, with their day- ahead and real-time markets, large geographies to aggregate diverse wind resources, large loads to aggregate with wind, large generation pools that tap conventional generator flexibility... offer the best environments for wind generation to develop." (Milligan et al. 2009b).

[64] This includes introducing sub-hourly markets that allow systems faster response to variability. Large ISOs with 5-minute markets typically have substantially lower wind integration costs.

[65] This could include scheduling generators over shorter time periods, and using sub-hourly wind forecasts instead of "persistence" forecasts that may actually contribute to short-term scheduling errors and increase regulation requirements.

[66] "Extensive changes will be required in the type of new generation built in the state: new units must have greater operating flexibility to start up and shut down without long delays; they must be able to operate at lower minimum loading levels; and they must have faster

ramping capability and regulation capability" (CAISO 2007). Examples of more responsive generators include certain aeroderivative gas turbines and reciprocating engines. For additional discussion of flexible generators, see Northwest Power and Conservation Council 2009.

[67] As an example, ERCOT currently obtains half of its spinning reserve requirements from responsive load, and the ERCOT event of February 2008 discussed previously is an example of an application of load as a source of "up" ramping, used to meet demand until conventional generators could be started and dispatched.

[68] In effect, combining individual VG and storage is essentially the creation of very small balancing areas. This is actually the opposite of how the grid is evolving, with the creation of larger balancing areas and the use of reserve sharing agreements across utilities. Balancing individual VG would dramatically increase reserve requirements and would incur much greater costs than at the system level.

[69] Use of dedicated long-distance transmission for wind or solar will be limited by the relatively low capacity factor of the resource. Storage could increase line-loading and help reduce curtailment due to transmission constraints. For additional discussion, see Denholm and Sioshansi 2009.

[70] This often refers to the time it takes after a power failure for an isolated system to switch from the grid to a backup generator. However, this term may also be useful to describe the ability to address forecast errors and bring up standby generators during times of unforeseen decreases in wind or conventional generation.

[71] This chart represents technologies actually deployed or proposed as of November 2008. It does not include a number of pre-commercial products or represent the total range of applications. For example, most of the batteries listed could be scaled up in either energy or power capacity, while at least 1 CAES plant of greater than 1,000 MW has been proposed. Alternatively, PSH plants of less than 50 MW have been constructed (ESA 2009).

[72] This excludes small distributed applications including uninterruptible power supplies, off-grid homes, and the substation batteries. These applications are dominated by lead-acid batteries.

[73] 100 MW is equivalent to a small power plant and negligible in terms of overall grid-scale capacity. As of January 2008, the United States had 1,087,791 MW of installed capacity (EIA 2009a).

[74] The U.S. electric grid uses alternative current (AC) while batteries, capacitors, and several other electric storage technologies charge and discharge direct current (DC).

[75] Furthermore, PHS and CAES depend on site-specific geologic conditions, which makes costs difficult to generalize.

[76] These devices have several names such as ultracapacitors and supercapacitors. "There is some uncertainty within the industry on the exact name for capacitors with massive storage capability. This is in part due to the many names of products by different manufacturers, but also due to the relative newness of the industry and recent advances." (EPRI 2003).

[77] A complete list – including capacity, location, date of initial operation, and ownership – is available from EIA Form EIA-860, "Annual Electric Generator Report."

[78] About 30 GW of new pumped hydro capacity has been proposed between 2006 and 2009. This represents more than double the existing capacity, and certainly implies there are considerable opportunities for new pumped hydro capacity (Adamson 2009).

[79] Losses in the storage system are relatively small and occur through heat exchange from the stored cold energy and the surrounding environment. These losses can be partially offset by the potential increase in compressor efficiency when making ice or chilled water in the cooler evening compared to the daytime, which achieves net efficiencies close to 100%.

[80] This conclusion depends on the anticipated cycle life and cost of EV batteries. See Sioshansi and Denholm 2009 and Peterson et al. 2010 for a discussion of the impact of battery life and cycling on the value of V2G. However, controlled charging (without V2G) is still a potentially significant source of flexibility, with the ability to raise the minimum load and avoid curtailment..

In: Energy Storage: Issues and Applications ISBN: 978-1-61209-517-2
Editors: Jonathan M. Bowen

Chapter 2

LIFECYCLE COST ANALYSIS OF HYDROGEN VERSUS OTHER TECHNOLOGIES FOR ELECTRICAL ENERGY STORAGE*

D. Steward, G. Saur, M. Penev and T. Ramsden

EXECUTIVE SUMMARY

As renewable electricity becomes a larger portion of the electricity generation mix, new strategies will be required to accommodate fluctuations in energy generation from these sources. One of the primary strategies proposed for integrating large amounts of renewable energy is using energy storage to absorb excess electricity-generating capacity during times of low demand and/or high rates of generation by renewable sources and then reconverting this stored energy into electricity during periods of high demand and/or low renewable generation.

Various energy storage technologies have been developed or proposed. The goal of this analysis was to develop a cost survey of the most-promising and/or mature energy storage technologies and compare them with several configurations employing hydrogen as the energy carrier. A simple energy arbitrage scenario was developed for a mid-sized energy storage system consisting of a 300-MWh nominal storage capacity that is charged during off-

* This is an edited, reformatted and augmented edition of a National Renewable Energy Laboratory publication, Report NREL/TP-560-46719, dated November 2009.

peak hours (18 hours per day on weekdays and all day on weekends) and discharged at a rate of 50 MW for 6 peak hours on weekdays.

For all the hydrogen cases, off-peak and/or excess renewable electricity is used to electrolyze water to produce hydrogen, which is stored in compressed gas tanks or underground geologic formations. The hydrogen is reconverted into electricity using a polymer electrolyte membrane (PEM) fuel cell or hydrogen expansion combustion turbine. The hydrogen storage scenarios are compared with the use of several battery systems (nickel cadmium, sodium sulfur, and vanadium redox), pumped hydro, and compressed air energy storage (CAES).

All the energy storage systems are evaluated for the same energy arbitrage scenario using consistent financial and operational assumptions. Costs and performance parameters for the technologies were gathered from literature sources and, in the case of the hydrogen expansion combustion turbine, Aspen Plus modeling. Producing excess hydrogen for use in vehicles or backup power is also evaluated. Two production levels are analyzed: 1,400 kg/day (roughly equivalent to the U.S. Department of Energy's standard model for small- scale distributed hydrogen production) and 12,000 kg/day. As for the purely energy arbitrage scenarios, it is assumed that hydrogen would be produced with offpeak/renewable electricity. Cost results for the analysis are presented in terms of the annualized ("levelized"[1]) cost for producing the energy output from the storage system: electricity fed back onto the grid during peak hours ($/kWh) and, in the case of producing excess hydrogen for vehicles, hydrogen ($/kg).

Figure ES – 1. summarizes the comparison of levelized cost of delivered electricity for hydrogen (green bars) and competing technologies (blue bars). For each technology, high-cost, mid-range, and low-cost cases were analyzed, and sensitivity analyses were performed to generate a range of possible costs for each case. In Figure ES - 1, the bottom of the bars represents the low end of the range for the low-cost cases, and the top of the bars represents the high end of the range for the high-cost cases. The numerals shown are the nominal values of the mid-range cases; these mid-range values do not represent a statistical determination of most-probable costs.

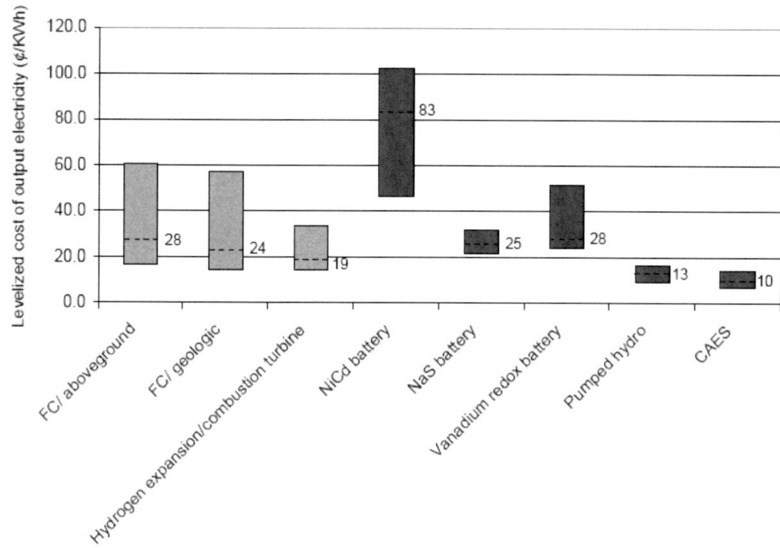

Figure ES- 1. Ranges of levelized cost of output electricity for electricity storage systems

The range of costs for each system reflects the range found in the literature and estimates of potential cost reductions as technologies develop. The hydrogen fuel cell scenario cost range reflects the comparative immaturity of fuel cell technologies for this application. It is anticipated that costs for fuel cells will decrease as the technology matures. Hydrogen combustion turbines could prove to be viable for energy storage applications and could provide additional flexibility to utilities through co-firing of mixtures of natural gas and hydrogen. Hydrogen technologies are competitive with battery systems for this application and could be a viable alternative to pumped hydro and CAES at sites where those technologies are not feasible.

Using hydrogen for energy storage provides unique opportunities for integration between the transportation and power sectors. An analysis was performed to evaluate the potential cost implications of producing excess hydrogen for vehicles in addition to what is needed for the electricity storage scenario. Producing a small amount of excess hydrogen (five 280-kg tanker-truck loads or 1,400 kg per day) for the vehicle market was evaluated by adding a hydrogen load to the mid-range energy storage case with aboveground storage of hydrogen. Producing this small amount of excess hydrogen reduces the overall levelized cost of energy for this scenario by about 6% compared with the purely energy arbitrage scenario.[2] The excess hydrogen is produced for $4.69/kg. Excess hydrogen produced in this way is still not competitive with hydrogen produced in a dedicated, distributed electrolysis process with the same daily output of hydrogen ($4.00/kg untaxed). However, for producing larger volumes of excess hydrogen to feed into a hydrogen pipeline, the scenario with energy storage plus excess hydrogen could be competitive with a dedicated hydrogen production facility. The energy storage plus excess hydrogen scenario produces 500 kg/hour (12,000 kg/day) of excess hydrogen for $3.33/kg (untaxed). A dedicated, centralized, 500-kg/hour electrolysis facility produces hydrogen for $6.86/kg (untaxed).

ACROYNMS AND ABBREVIATIONS

AC	alternating current
BoP	balance of plant
CAES	compressed air energy storage
DC	direct current
DOE	U.S. Department of Energy
EPRI	Electric Power Research Institute
FC	fuel cell
H_2	hydrogen
ICE	internal combustion engine
IGCC	integrated gasification combined cycle
KOH	potassium hydroxide
LCOE	levelized cost of energy
LHV	lower heating value
LPG	liquefied petroleum gas
MP	mass production
MYPP	Multi-Year Research, Development, and Demonstration Plan
NaS	sodium sulfur
NiCd	nickel cadmium
NO_x	nitrogen oxides
NPC	net present cost
O&M	operating and maintenance
PCS	power conditioning system
PEM	polymer electrolyte membrane
VR	vanadium redox

1. Introduction

This chapter presents the results of an analysis evaluating the economic viability of hydrogen for medium- to large-scale electrical energy storage applications compared with three other storage technologies.

- Batteries
- Pumped hydro
- Compressed air energy storage (CAES)

2. Analysis Methodology

Potentially viable hydrogen production and storage scenarios were developed, and a lifecycle economic analysis was performed to determine the levelized cost of delivering energy for these scenarios. The results were benchmarked against the three competing technologies on an "apples to apples" basis. Sensitivity analyses were performed for three cost cases for each technology.

2.1. Scenario Definitions

An energy arbitrage scenario was developed to simulate the storage and dispatch back to the grid of electricity at favorable cost, supply, and demand conditions. Additional scenarios were developed to simulate the production of excess hydrogen along with the energy storage and dispatch functions.

2.1.1. Energy Arbitrage Scenario

Although each utility service area is unique, analysis of a utility's hour-by-hour marginal costs generally shows that a percentage of electricity demand is met at very low cost— typically \$0.02–\$0.03 per kWh or less— using baseload-generation units. Above that, the marginal production cost from peak-generation units increases to \$0.04–\$0.10 or more per kWh (Ramsden et al. 2008). Systems that store electricity during low-demand/cost periods and supply it during high-demand/cost periods displace this higher-marginal-cost peaking power. A study by Sioshansi et al. (2008) for the PJM system[3] showed that, for small energy storage devices for which the electricity produced for energy storage does not affect grid electricity prices, more than 50% of the total capturable arbitrage value of the system is obtained during the first four hours of discharge. With longer-term storage allowing inter-day arbitrage and weekend charging, eight hours of storage captures about 85% of the potential value. Six hours of discharge was selected for this study.

An energy arbitrage scenario matching the scenario in Ramsden et al. (2008) was used in this study in which grid electricity from renewable or other sources is stored when electricity demand and cost are low and/or when renewable sources must be used in the absence of equivalent demand. The stored energy is converted back to electricity and dispatched to the grid during periods of peak electricity demand and cost. The scenario corresponds to the bulk energy storage application definition used by Schoenung and Has senzahl (2003). Table 1 lists key parameters of this scenario.

The primary figure of merit for this scenario is levelized cost of delivered electricity (LCOE), defined as the total annualized cost of the storage system divided by the annual energy output from the system. In the case of the purely energy arbitrage scenario, the output energy is the yearly total of electricity fed onto the grid during peak hours on weekdays. For the scenarios including production of excess hydrogen for the vehicle market, the output energy for the system is the yearly electricity production plus the yearly production of hydrogen used for vehicles. The annualized costs include the initial capital investment, replacement costs, end-of-life costs or credits, and fixed and variable operating costs summed over the lifetime of the facility and divided by the facility lifespan. The storage system might also meet requirements for spinning reserve and other services, but no value is assigned to these services because of the wide variety of factors affecting the value of these services for a particular utility and the variations in requirements for different utilities. Off-peak electricity costs used in this study do not include transmission and distribution charges. The analysis also does not include consideration of business taxes.

2.1.2. Excess Hydrogen for the Vehicle Market

Using hydrogen for large-scale energy storage could also provide an economical source of hydrogen for fuel cell vehicles. Two basic scenarios for production of excess hydrogen for vehicles were evaluated. In both scenarios, the energy storage scenario remains unchanged, but additional hydrogen is produced for use in vehicles.

Table 1. Key Parameters of Energy Arbitrage Scenario

Energy storage system capacity	50 MW for 6 peak hours each weekday (300 MWh/day)[1]
Plant life	40 years[2]
Interest rate on debt	10%[3]
Debt financing	100%[3]
Off-peak/renewable electricity cost	\$0.025–\$0.06/kWh; \$0.038/kWh is used as the baselinerenewable/off-peak electricity price for this study[4]
Natural gas cost (CAES system)	\$7/MMBtu

[1] The remaining 18 hours per day on weekdays and all day on weekends are assumed to be off-peak hours when the system can be charged.

[2] Some equipment is replaced at more frequent intervals.

[3] 100% debt financing and 10% interest rate is equivalent to the H2A model standard assumption of 100% equity financing and 10% IRR.

[4] \$0.038/kWh is the value used for wind-generated electricity in Ramsden et al. (2008) and is consistent with current estimates for wind-generated electricity costs. \$0.06/kWh represents a conservative cost for electricity from renewable sources. \$0.025/kWh represents very-low-cost, off-peak electricity for sensitivity analyses.

In the first scenario, it is assumed there would be demand for five 280-kg (1,400-kg/day) gaseous tankers of hydrogen per day from the storage system. This scenario produces 511,000 kg of hydrogen per year for vehicle use, which is roughly equivalent to the output from one forecourt hydrogen station and about 12% of the hydrogen produced for the energy storage scenario.[4] It is assumed that the electrolyzer system would be sized slightly larger but would operate on the same schedule as the energy storage scenarios to take advantage of low electricity prices and provide additional load during times of low electricity demand. It is assumed that the hydrogen would be stored in aboveground steel tanks that could be loaded onto a trailer.

The second scenario is similar to the first in that the electrolyzer system is sized to accommodate production of additional hydrogen during off-peak hours. However, in this scenario, approximately the same amount of hydrogen is produced for vehicles on a yearly basis as is produced to fuel the fuel cell for peak electricity production. This scenario assumes that 500 kg/hour of hydrogen flows into a pipeline at all times. Geologic storage and pipeline transport of the hydrogen is assumed. Costs were developed for production and storage of the hydrogen for both scenarios. Costs for transport of the hydrogen to refueling stations were not considered.

2.2. HOMER Model

The facility lifecycle economic analyses were performed using HOMER, the National Renewable Energy Laboratory's distributed-generation economic model.[5] System components, available energy resources, and loads were modeled hour by hour for a single year with energy flows and costs held constant over a given hour. HOMER requires inputs such as subsystem component options and performance, capital and replacement costs, fuel and electricity costs, and resource availability. It uses these inputs to simulate different system configurations and generate a list of feasible configurations sorted by net present cost (NPC). HOMER also reports the cost of energy produced for each feasible system configuration, reported on a $/kWh basis. Because the energy storage systems modeled in this study always produce the same amount of electricity, the system configuration with the lowest NPC is also the configuration with the lowest cost of electricity. As such, the system configuration with the lowest cost of energy is also the most economic solution.

HOMER defines LCOE as the average cost per kWh of useful electrical energy (and hydrogen if produced as a product rather than as a storage medium) produced by the system. To calculate the LCOE, HOMER divides the annualized cost of producing electricity and/or hydrogen (the total annualized cost minus the cost of serving the thermal load) by the total useful electric and hydrogen energy production. For the energy arbitrage system modeled in this study, the cost of electricity is simply the total annualized cost of the system ($/yr) divided by the total primary AC load served (kWh/yr). Details on the calculations can be found in Lambert et al. (2006).

2.3. Cost and Sensitivity Analyses

The energy arbitrage scenario was analyzed for four energy storage technologies and three cost values:

- Hydrogen (fuel cell production of electricity and hydrogen combustion turbine)
- Batteries (vanadium redox, nickel cadmium, and sodium sulfur)
- Pumped hydro
- CAES

High cost. These cost and efficiency values represent a conservative estimate for current technologies. In most cases, the values were derived from studies that compiled costs from actual installations. In many cases, these costs represent first-generation installations and conservative estimates of costs for installations that could be built at the time the referenced study was developed.

Mid-range cost. These values were derived from estimated technology improvement, size scale-up, and bulk manufacturing cost reductions projected to be available in the near future (3 to 5 years).

Low cost. These values were derived assuming optimal or "fully mature" technologies and many large-scale installations. In some cases, they were derived from existing studies, and in some cases they were based on percentage reductions in cost that are consistent with projections for similar technologies.

Pumped hydro is the most mature technology considered in this analysis. Pumped hydro systems of varying sizes, including very large installations up to nearly 3,000 MW, have been installed throughout the world. The remaining technologies included in this analysis are much less established. Although CAES involves well-established commercial technologies (compressors and gas turbines), and many designs and projects have been proposed, only two facilities have been built. Battery systems are also commercially available but have not been implemented extensively for bulk energy storage. Electrolyzers and hydrogen fuel cells are also commercially available but have never been combined for bulk energy storage applications. The timeframe for achieving the low-cost case may be much greater for the fuel cell systems than for more established technologies.

Costs for energy storage systems depend on the power (kW) and energy (kWh) capacity of the systems in addition to fixed costs that are independent of system size (EPRI 2007). Costs that depend on system power capacity include the power conditioning system (PCS), cell stack for fuel cells and flow batteries, and pumping or compression equipment for pumped hydro and CAES systems. Costs that depend on energy storage capacity include battery capital costs, some balance of plant (BoP), and electrolyte and electrolyte storage systems for vanadium redox batteries. Fixed costs include, for example, control systems, construction management, and permitting. The total capital cost for the system can be described by the following equation (EPRI 2007):

$$\text{Capital Cost} = (\text{power capacity} \times \$/\text{kW}) + (\text{duration} \times \$/\text{kWh}) + \text{fixed cost} \qquad \text{(Eq. 1)}$$

Although this equation generally holds, capital cost components are represented differently in different references used in this analysis. To maintain consistency and allow for comparisons between different sources, values in this chapter are presented using the breakdown shown in Equation 1. These may differ from the presentation in the original source. Cost values in tables presenting an overview of sources and background information are shown in their original units. All values used in the cost modeling for this study were escalated to $2008 using the GDP Implicit Deflator Price Index.[6]

The storage system must be sized to account for energy losses during the electricity conversion and storage processes as well as equipment mechanical or voltage limitations that prevent the system from being discharged fully during operation (Schoenung and Hassenzahl 2003). Cost modeling must account for the oversizing of the system needed to provide the required delivered electricity. Voltage/capacity limitations are especially important for sizing of battery systems. The HOMER battery model accounts for these limitations based on default or user-input battery profiles. Cost information is typically presented based on the net power capacity per discharge cycle and net power output of the plant. Therefore, all capacity-related costs ($/kWh) are based on 300 MWh (50 MW × 6 hours), and all power-related costs ($/kW) are based on 50-MW net power output for the plant.

Lifecycle costs for each of the technology cases were calculated using HOMER. Results are presented for initial capital investment and system NPC, including operating costs. Low-cost, mid-range, and high-cost cases were developed for each technology.

Table 2. Values for Energy Storage Technology Sensitivity Analyses

	Value	Notes
Initial capital	± 20 %	Battery systems, electrolyzer/fuel cell systems, hydrogen aboveground tanks, compressors, and pumps
Fixed operating cost	± 20 %	
Off-peak electricity cost	\$0.025, \$0.038, and \$0.06 per kWh	The price of off-peak electricity varies significantly with location and local utilityrate structures. The price of off-peak electricity may vary from \$0.02/kWh to \$0.14/kWh (EPRI 2006).
Cavern/reservoir cost	± 50 %	Sensitivity values were varied over a wider range owing to higher variability incosts due to location constraints.
Vanadium redox electrolyte	± 50 %	Sensitivity values were varied over a wider range owing to high volatility in vanadium prices.

Sensitivity analyses were performed for each of the three cost cases for all the technologies. Sensitivity analyses include initial capital costs, cost of electricity for charging the system, and fixed operating costs. Additional sensitivity analyses were performed for the cavern/tank/reservoir costs for CAES, fuel cell, and pumped hydro systems and the CAES system cost sensitivity to the price of natural gas fuel for the combustion turbine. Table 2 lists the sensitivity values used in the analysis.

3. Energy Storage Technologies and Costs

The energy storage technologies analyzed in this study range in maturity from novel concepts to commercial systems used for decades. Current cost and performance information was combined with projections of future technological development to model each technology and cost case.

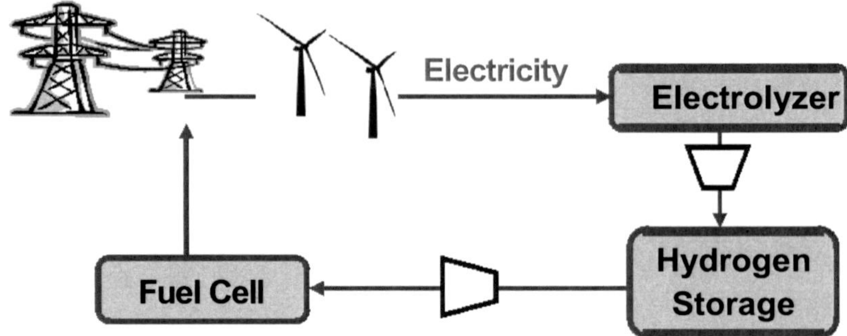

Figure 1. Hydrogen electrical energy storage and dispatch scenario

3.1. Hydrogen Systems

The energy arbitrage scenario was analyzed for three hydrogen cases. In each case, an electrolyzer system converts electricity into hydrogen (by electrolyzing water), which is stored in either steel tanks or geologic caverns for later conversion back into electricity. In the first two cases, a fuel cell system converts the hydrogen into electricity (Figure 1). In the third case, the hydrogen fuels an expansion/combustion turbine system.

Ramsden et al. (2008) evaluated use of a hydrogen internal combustion engine (ICE) for production of electricity from the storage system. They found that the low roundtrip efficiency of between 21% and 36% for the ICE cases compared with 36% to 50% roundtrip efficiency for the fuel cell cases did not compensate for the potentially lower cost of the ICE. The initial capital cost of the ICE system was, in fact, higher than the initial capital costs for the comparable fuel cell scenario that was also based on steel tank (aboveground) storage. The capital cost for the hydrogen ICE was only slightly less than the cost for the fuel cell: \$720/kW (\$2005) for the ICE versus \$740/kW (\$2005) for the fuel cell for the near-term case. The low efficiency of the hydrogen ICE required both increased hydrogen storage capacity and a larger electrolyzer system, which increased the overall cost. Therefore, a fuel cell was used for electricity generation from the storage system in most cases evaluated in this study. However, the use of hydrogen in place of natural gas as the fuel for the combustion turbine in a compressed air/compressed hydrogen comparison was analyzed (see Section 3.1.3).

In summary, the three hydrogen energy arbitrage scenarios are as follows:

- Case 1—Compressed hydrogen stored aboveground in steel tanks and polymer electrolyte membrane (PEM, also called proton exchange membrane) fuel cell for conversion of the stored hydrogen energy into electricity.
- Case 2—Compressed hydrogen stored underground (i.e., geologic storage) in salt caverns and PEM fuel cell for conversion of the stored hydrogen energy into electricity.
- Case 3—Compressed hydrogen stored underground (i.e., geologic storage) in salt caverns and hydrogen combustion turbine for conversion of hydrogen energy into electricity.

The use of hydrogen for energy storage provides unique opportunities for integration between the transportation and power sectors. Additional analyses were performed to evaluate the potential cost implications of producing excess hydrogen for use in vehicles, in addition to what is needed for the electricity storage scenario. See Section 2.1.2 for a description of these scenarios.

Geologic Storage

In general, geologic hydrogen storage is anticipated to be considerably cheaper than storing hydrogen in steel tanks. However, development of geologic storage reservoirs is highly dependant on the characteristics of the geologic formation. EPRI-DOE (2003) cost estimates for developing underground storage facilities for CAES systems range from $0.10/kWh ($2003) for porous rock formations to $30/kWh ($2003) for excavation of hard-rock formations. The storage volume required for a hydrogen-based system would be orders of magnitude less than the volume required for an equivalent-energy-capacity CAES reservoir because of the higher caloric value of the hydrogen. Crotogino and Huebner (2008) estimated the energy density for a typical CAES system at 2.4 kWh/m^3; for a comparable hydrogen reservoir, they estimated 170 kWh/m^3. Table 3 shows the range of costs for geologic storage cavern development for CAES and hydrogen assuming the energy density values given by Crotogino and Huebner. Geologic storage costs for hydrogen developed for the H2A Delivery Components Model (Argonne National Laboratory 2009/2009a) are shown for comparison. That analysis is based on a natural gas storage facility in Saltville Virginia.[7] The values given in the H2A Delivery Components Model are used for the geologic storage costs in this chapter. The current cost estimate for storage of hydrogen in aboveground tanks is $623/kg or ~$ 19/kWh (Ramsden et al. 2008).

Table 3. Costs of Geologic Storage Cavern Development for CAES and Hydrogen

Formation Type	Air $/kWh ($2003)	Air $/kWh ($2008)	Air $/m³($2008)	Hydrogen $/kWh[1]
Solution-mined salt caverns[2]	1.00	1.20	2.88	0.02
Dry-mined salt caverns[2]	10.00	11.50	27.60	0.16
Rock caverns created by excavating comparatively impervious rock formations[2]	30.00	35.00	84.00	0.49
Naturally occurring porous rock formations (e.g., sandstone and fissured limestone) from depleted gas or oilfields[2]	0.10	0.12	0.29	0.002
Abandoned limestone or coalmines[2]	10.00	11.50	27.60	0.16
Geologic storage of hydrogen3	N/A	N/A	N/A	0.30

[1] Hydrogen storage cavern development cost is calculated assuming the same $\$/m^3$ as for CAES cavern development and energy density from Crotogino and Huebner (2008).

[2] Source: EPRI (2003) and Crotogino and Huebner (2008).

[3] Equation from H2A Delivery Scenario Analysis Model Version 2.02, for 41,000-kg usable storage capacity, www.hydrogen delivery.html.

Hydrogen storage in geologic formations would have considerable advantages over comparable air storage. However, systems using geologic storage can be more difficult to site compared with steel tank systems. Hydrogen is currently stored in solution-mined salt domes in the U.S. Gulf Coast. Figure 2 shows the location of known U.S. salt deposits. A major advantage of solution-mined salt caverns is that they can be developed to any size, allowing use of minimal volumes of cushion gas (the gas required to maintain the minimum pressure in the cavern) even for very small installations such as the one envisioned in this study.

Figure 2. Known U.S. salt deposits (Casey 2009)

Potential issues involved with geologic storage of hydrogen include salt flow over time (shrinkage of approximately 0.2 5% per year), subsidence, and deformation/breakage of the shaft causing equipment damage (Casey 2009). Alternatives include hydrogen storage, as liquid or as high-pressure gas, in buried tanks, or in hard-rock formations using a water curtain to contain the hydrogen (IKA 2009). Hydrogen stored in caverns may also require purification before it can be used in PEM fuel cells (in stationary or vehicle applications).

Figure 3. Location of major oil and gas fields in the United States (Borns and Lord 2008)

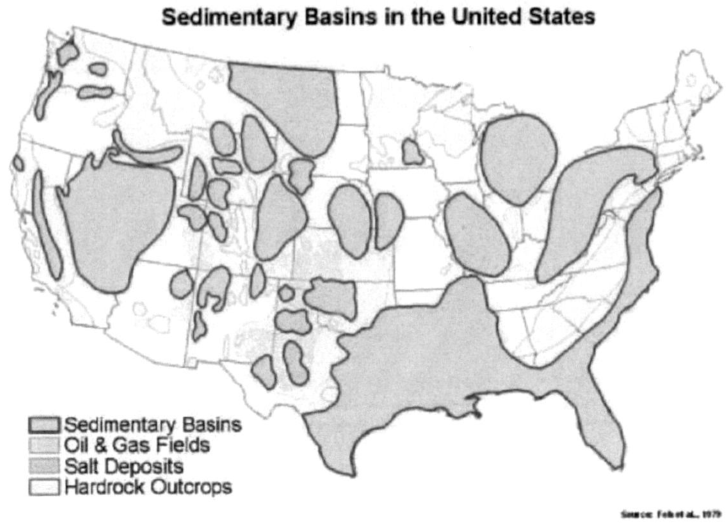

Figure 4. Location of sedimentary basins in the United States (Borns and Lord 2008)

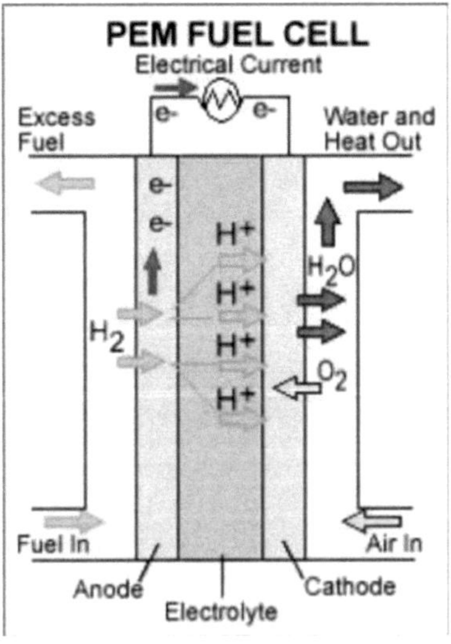

Figure 5. Schematic of a PEM fuel cell (DOE 2009)

Additional types of geologic formations may be suitable for hydrogen storage. The U.S. Department of Energy (DOE) Fuel Cell Technologies Program has investigated the potential for hydrogen storage in geologic formations and identified areas for further research (Borns and Lord 2008). In general, caverns, whether natural or mined, must provide containment of the gas. In excavated caverns, this can be accomplished by lining the cavern with steel or using hydraulic pressure in the surrounding rock. In addition to salt caverns, depleted gas reservoirs and aquifers were identified as probable candidates for underground storage of hydrogen. Figure 3 and Figure 4 show locations of depleted gas reservoirs and aquifers in the United States. Further research is needed to ensure hydrogen containment and purity, including investigation of hydrogen mobility in different rock types, hydrogen embrittlement, gas mixing, and the effect of hydrogen on rock properties.

3.1.1. Hydrogen Fuel Cell System Description

Stationary PEM fuel cells were selected for the energy arbitrage scenario examined in this study (Figure 5). PEM fuel cells may be more appropriate for the study application because of their low operating temperature and ability to cycle on and off more readily than fuel cells that operate at higher temperatures. However, the scenario envisioned here would require much larger fuel cells than are currently available. A literature search was performed to identify cost and efficiency trends for PEM fuel cells. Values used in this analysis were derived primarily from the references shown in Table 4.

Figure 6 illustrates the spread in costs for existing stationary PEM fuel cell installations at military bases and stationary PEM fuel cell cost estimates for mass-produced fuel cells (Dhathathreyan and Rajalakshmi 2007, Stone 2005, Lipman et al. 2004). The year is the installation year for the fuel cells installed at military bases and the year referenced in the studies. These costs are presented to illustrate the range in values for cost estimates and actual costs for installations of varying size and purpose. The fuel cells installed at the military bases were very small (5-kW) systems. No attempt was made to convert the dollars to a reference year.

Table 4. References Used to Define PEM Cost and Efficiency Values

High Cost	Mid-range	Low Cost
Lipman et al. (2004)	Stone (2005)	Lipman et al. (2004)
LoganEnergy (2008)[1]	DOE (2007)	DOE (2007)
LoganEnergy (2007)[1]		
Oak Ridge (2008)		

[1] Plug Power 5-kW PEM fuel cell for backup power, bottled hydrogen supply

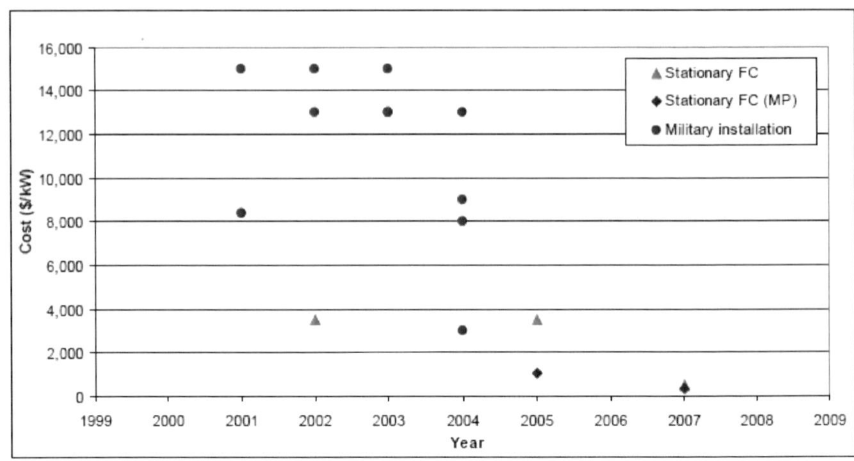

Figure 6. Existing PEM fuel cell costs and estimates for mass production (MP) of PEM fuel cells

Table 5 lists the specifications for the PEM fuel cell used in this analysis. Current-timeframe costs presented for the PEM fuel cells are for the fuel cell stack and auxiliary BoP equipment taken from actual installations. Costs for a reformer/shift reactor are not included in these costs. PCS components for the electrolyzer and fuel cell are assumed to be included in the costs for these systems.

The efficiency values for the PEM fuel cell system are derived from DOE's Multi-Year Research, Development, and Demonstration Plan (MYPP) status and target values for stationary and transportation PEM fuel cell power systems (DOE 2007). The high-cost and mid-range cases are based on large (250 kW and larger) stationary fuel cells. The low-cost case assumes somewhat smaller (80–100 kW) fuel cells typically used in automotive applications. For assessing progress against target values, the MYPP defines electrical efficiency for stationary PEM fuel cell systems as "the ratio of DC output energy to the LHV [lower heating value] of the input fuel (natural gas or LPG [liquefied petroleum gas]) average value at rated power over life of power plant." The MYPP definition assumes that stationary hydrogen fuel cells will be integrated with a fuel reformer, which converts natural gas into a hydrogen-rich feed stream for the fuel cell. For the analysis of hydrogen energy storage scenarios, however, hydrogen is produced by an electrolyzer rather than by natural gas reforming. Therefore, efficiency values for the fuel cell without the reformer are used here. The MYPP shows an overall system efficiency (including the reformer) of 32% in 2005 with a target of 40% in 2011. Assuming an overall system efficiency of 40% and a reformer efficiency of 85% (Stevens and Lightner 2003) gives a fuel cell subsystem efficiency of 47%. The low-cost case efficiency value is based on DOE targets for automotive PEM fuel cells operating at 25% of rated power (DOE 2007).

For the purpose of this analysis, it was assumed that the electrolyzer and fuel subsystem costs include power conversion and BoP costs. Fuel cell costs are based on the references listed in Table 4. DOE estimates of current costs and near-term targets ($2,500/kW and $750/kW, respectively, in $2005) for stationary PEM fuel cells are comparable to the high-cost and mid-range cases (DOE 2007). The value of $434/kW for the low-cost case is consistent with projected values for stationary fuel cells (Lipman et al. 2004) as well as DOE automotive fuel cell cost targets adjusted to account for low-power operation and increased BoP costs associated with the larger number of fuel cells required. Durability is assumed consistent with projections for stationary fuel cells.

Table 5. Components for Analysis of PEM Fuel Cell

System Component	High-Cost Case Values	Mid-Range Case Values	Low-Cost CaseValues
Fuel cell system installed capital cost($2008)	$3,000/kW	$813/kW	$434/kW
Stack replacement frequency/cost	13 yr[1]/30% of initial capital cost	15 yr/30% of initial capital cost	26 yr[1]/30% of initial capital cost
O&M costs	$50/kW-yr[2]	$27/kW-yr	$20/kW-yr[2]
Fuel cell life	13 yr (20,000-hour operation)	15 yr (24,000-hour operation)	26 yr (40,000-hour operation)
Fuel cell system efficiency (LHV)	47%	53%[3]	58%[4]

[1] DOE (2007), Chapter 3.4; 20,000 hours for stationary PEM reformate system fuel cells 5–250 kW has been demonstrated. The goal for 2011 is, "By 2011, develop a distributed generation PEM fuel cell system operating on natural gas or LPG that achieves 40% electrical efficiency and 40,000 hours durability at $750/kW." Validated by 2014. Twenty thousand hours (13 years) was used for the high-cost value, and 40,000 hours (26 years) was used for the low-cost value.

[2] Values are from Lipman et al. (2004).

[3] Current technology value for stack efficiency is approximately 55% (O'Hayre et al. 2006). Value is mid-way between the high and low estimates.

[4] Assumed stack efficiency of 60% (MYPP 2010 target for direct hydrogen fuel cells for transportation) with 2% conversion losses for integrated system.

Table 6 shows electrolyzer and storage system efficiency and costs. The electrolyzer system is based on the H2A central electrolysis cost analysis models.[8] The system modeled is a standalone, grid-powered electrolyzer system based on the Hydro bipolar alkaline electrolyzer (Atmospheric Type No. 5040 - 5150 Amp DC). Each electrolyzer is capable of producing 1,049 kg/day of hydrogen. The electrolyzer units use process water for electrolysis and cooling water for cooling. Potassium hydroxide (KOH) is needed for the electrolyte in the system. The system includes the following equipment: transformer, thyristor, electrolyzer unit, lye tank, feed water demineralizer, hydrogen scrubber, gas holder, two compressor units, deoxidizer, and twin tower dryer.

Table 6. Hydrogen fuel cell system efficiency and cost

	High-Cost Values	**Mid-Range Values**	**Low-Cost Values**
Electrolyzer system efficiency (LHV/HHV)	62% / 73%	68.5% / 81%	73.5% / 87%
Electrolyzer capital cost ($2008)	$830/kW	$450/kW	$340/kW
Steel tank storage compressor electricity use (kWh/kg H_2)[1]	4	4	4
Steel tank storage capital cost ($2008)[2,3]	$51.2 million	$29.8 million	$16.9 million
Geologic storage compressor electricity use (kWh/kg H_2)[4]	2.2	2.2	2.2
Geologic storage capital cost ($2008)[5,6]	$4.75 million	$4.48 million	$4.44 million

[1] Hydrogen compression for tube trucks (05D_H2A_Delivery_Scenario_Analysis_Model_Version_2.02.xls).

[2] Capital cost includes tanks, compressor, and BoP.

[3] Hydrogen compressed to 2,500 psi.

[4] Hydrogen compression for bulk storage (05D_H2A_Delivery_Scenario_Analysis_Model_Version_2.02.xls).

[5] Hydrogen compressed to 1,800 psi.

[6] Cavern and compressor.

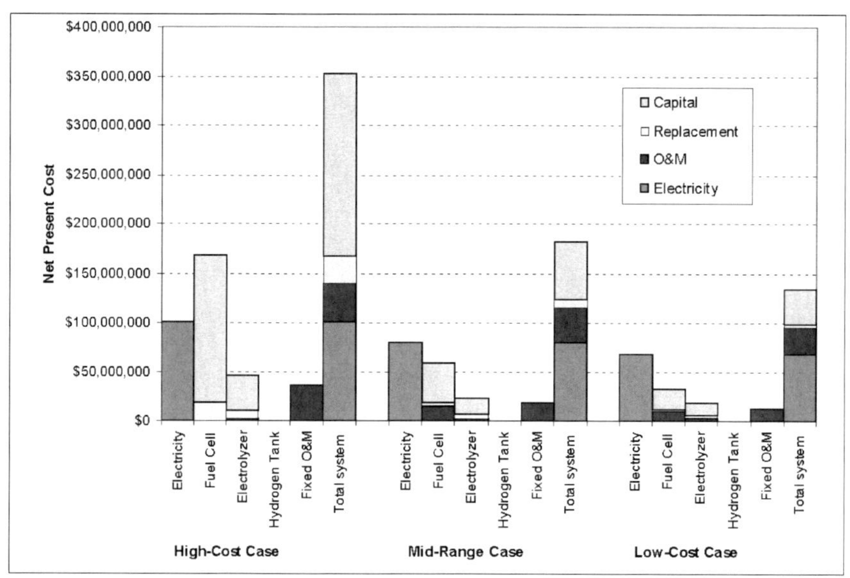

Figure 7. Hydrogen fuel cell energy arbitrage scenario with geologic storage NPC

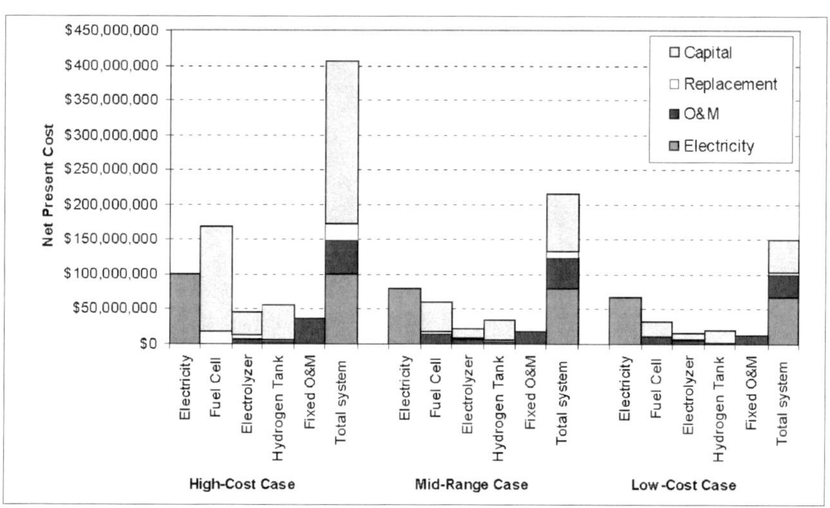

Figure 8. Hydrogen fuel cell energy arbitrage scenario with aboveground storage NPC

3.1.2. Hydrogen Fuel Cell System Cost Results

Figure 7 presents the NPC values for the hydrogen fuel cell energy arbitrage scenario with geologic hydrogen storage. For all the fuel cell cases, purchase of off-peak electricity constitutes a significant fraction of the overall cost. This is due to the low overall roundtrip efficiency of the systems. The roundtrip LHV efficiencies are 28%, 35%, and 41% for the high-cost, mid-range, and low-cost cases, respectively.

An additional case was run using the DOE cost target value of $45/kW for automotive PEM fuel cells, which is almost ten times lower than the $434/kW for the PEM fuel cells in the low-cost case (DOE 2007). The automotive fuel cell was assumed to have a 5,000- hour life, about one quarter to one eighth of the expected operating life for stationary PEM fuel cells. The total system NPC for the automotive fuel cell case is approximately $116 million, giving an LCOE of about $0.1 5/kWh. This compares with a NPC of $134 million and LCOE of about $0.1 8/kWh for the stationary fuel cell with geologic hydrogen storage low-cost case. The high replacement frequency for the automotive fuel cell reduces the advantage of the lower capital cost.

Figure 8 presents the NPC for the hydrogen fuel cell energy arbitrage scenario using aboveground steel tank storage for the hydrogen. As for the geologic storage fuel cell cases, purchase of off-peak electricity constitutes a significant fraction of the overall cost. The roundtrip LHV efficiencies are 27%, 34%, and 39% for the high-cost, mid-range, and low-cost cases, respectively.

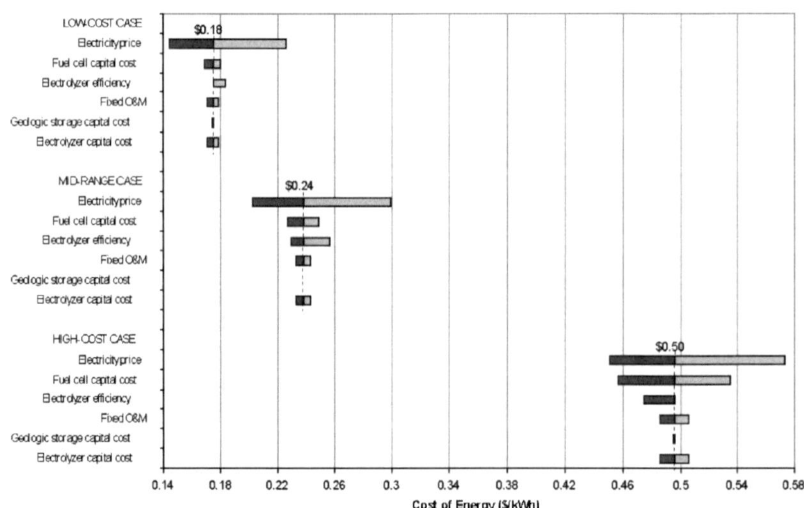

Figure 9. Cost sensitivity for hydrogen fuel cell scenario with geologic hydrogen storage

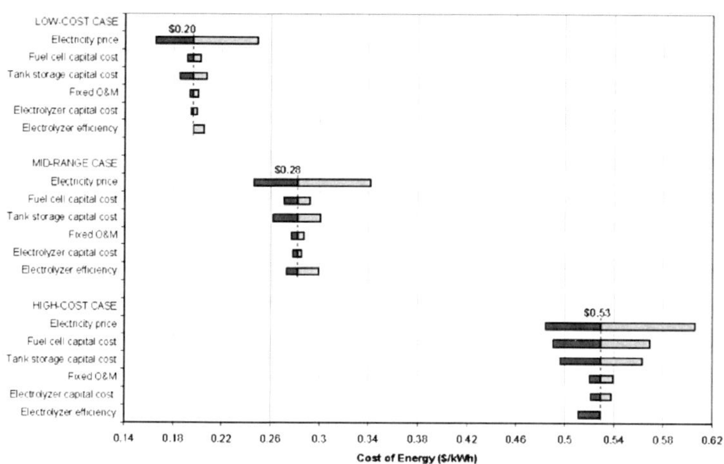

Figure 10. Cost sensitivity for hydrogen fuel cell scenario with steel tank hydrogen storage

Figure 9 and Figure 10 present the sensitivities for the hydrogen fuel cell energy arbitrage scenarios. For the fuel cell cases, the electrolyzer efficiency baseline value was varied for the three cost cases (see Table 6). These same values were used in the sensitivity analysis, resulting in "one-tailed" sensitivities for the low- and high-cost cases. In the mid-range and low-cost geologic storage cases, the electrolyzer efficiency is the second most important cost driver after off-peak electricity price. The overall cost impact of varying the geologic storage cost by ± 50% has a negligible effect on the overall cost of delivered energy for the geologic storage cases because of the very high energy density of hydrogen. Using aboveground storage rather than geologic storage adds between 6% and 18% to the cost of delivered energy.

3.1.3. Hydrogen Expansion Combustion Turbine System Description

Several companies are developing hydrogen combustion turbines. Toshiba is developing a hydrogen combustion turbine that generates steam from hydrogen and oxygen to run a steam turbine for electricity under the Japanese World Energy Network (WE-NET) research program. Pilavachi et al. (2009) list the efficiency for a hydrogen combustion turbine in the 70% range (LHV). The technology involves a combustion chamber in which hydrogen and oxygen combine to produce steam, which is then used to drive a turbine. In the energy storage scenario envisioned for this study, pure oxygen is collected from the electrolyzer and used for co-firing with the hydrogen in the plant. Costs for the hydrogen turbine using the efficiency stated in Pilavachi et al. (2009) represent the low- cost case. Collection, compression, and storage of the oxygen are assumed to be included in the costs. In 2005, General Electric and Siemens were awarded DOE research grants for hydrogen gas turbine development as part of advanced hydrogen turbine research for use in integrated gasification combined cycle (IGCC) plants (DOE-FE 2005).

Costs shown in Table 7 were collected for gas turbine power plants that are part of a larger system. Costs include equipment, installation, auxiliaries, and BoP. Costs for hydrogen gas turbines are generally seen in the context of IGCC or biomass IGCC plants.

Table 7. Costs for Conventional and Hydrogen-Fueled Gas Turbine Plants

Source	Year	Raw data	Converted $2008/kW	Notes
Afgan and Carvalho (2004)	2004	750 €/kW	$1,044	From Onanda.com historical data, using average Euro:USD for 2004 = 1.244; based on simple natural gas turbine plant
Phadke et al. (2008)	2008	$758/kW	$758	Compares several coal cycles;this isplant for IGCC gas turbine.
Siemens (2007)	2008	< $,1000	$1,000	"Power block (equipment + construction): 2 hydrogen-fueled gas turbines, 2 heat recovery steam generators, 1 steam turbine, 3 generators, and all associated auxiliaries/controls/BoP equipment"
Pilavachi et al. (2009)	2008	680 €/kW	$1,001	From Onanda.com historical data, using average Euro:USD for 2008 = 1.47; costs includes total power plantcosts - equipment and installation

A hydrogen expansion combustion turbine was substituted for the PEM fuel cell in two cases, one using geologic storage and one using aboveground tank storage. Costs for the electrolyzer, compressor, and storage system are the same as for the mid-range cost fuel cell cases. Aspen modeling was performed to derive the energy balance information for the hydrogen combustion turbine cases. A process flow diagram for the geologic storage case is shown in Figure 11. Hydrogen is compressed to a maximum cavern pressure of 1,262 psia using a four-stage compressor with intercoolers.

For the power generation system, hydrogen from the storage system is preheated with combustion exhaust prior to the expander stage. Compression requires approximately 0.76 kWh/kg of hydrogen. Approximately 0.83 kWh/kg is recovered from the preheated hydrogen in the expander. For the cost analysis, it is assumed that additional electricity is purchased for compression and that compression energy is recovered. The combustion turbine efficiency values given in Table 8 include additional energy recovered from expansion. Combustion air is fed to the turbine at a stoichiometric ratio of 3.7. The high airflow provides cooling, which reduces nitrogen oxides (NO_x) emissions (Dennis 2008a, Dennis 2008b, Juste 2006). NO_x emissions could be reduced further using a catalytic process.

Table 8. Components for Analysis of Hydrogen Combustion Turbine

System Component[1]	High-Cost Values (aboveground storage)	Mid-Range Values (geologic storage)	Low-Cost Values (geologic storage)
Expansion combustion turbine	$1 ,000/kW	$1 ,000/kW	$760/kW
Replacement frequency/cost	20 years / 100% of initial capital cost	20 years / 100% of initial capital cost	20 years / 100% of initial capital cost
Turbine O&M costs	$0.008/kW-yr	$0.008/kW-yr	$0.006/kW-yr
Combustion turbine efficiency[2]	42%	70%	70%

[1] Pilavachi et al. (2009). Costs are presented in Euros. Values are assumed to be in €2009. A conversion rate of $1.47/€ was assumed.

[2] Pilavachi et al. (2009).

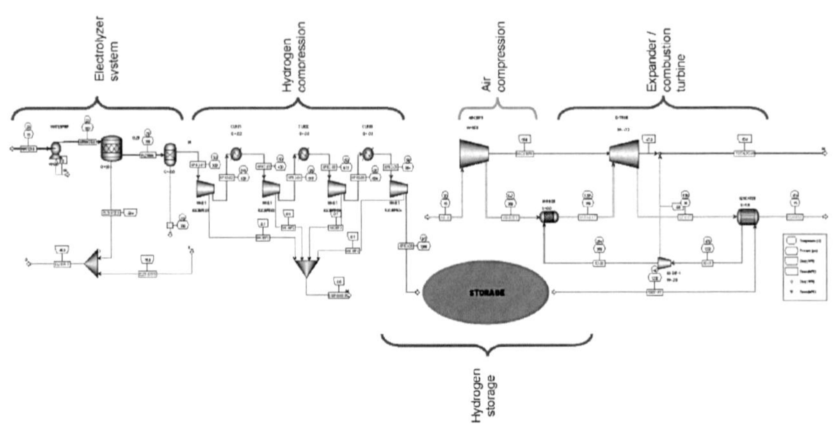

Figure 11. Process flow diagram for hydrogen combustion turbine system

Approximately 58% of the turbine output power is consumed for compression of the air for the modeled combustion turbine. The energy efficiency of the combustion turbine modeled in Aspen was used for the high-cost case. Plant cost values from Phadke et al. (2008) were used for the low-cost case. Cost values for the analysis of the hydrogen combustion turbine are given in Table 8.

3.1.4. Hydrogen Expansion Combustion Turbine System Cost Results

Figure 12 presents the NPC for the hydrogen expansion combustion turbine energy arbitrage scenario using aboveground steel tank storage for the high-cost case and geologic storage for the mid-range and low-cost cases. The roundtrip (AC-to-AC) efficiency for the low-cost and mid-range cases is 48% (LHV). The roundtrip efficiency for the high-cost case is 29% (LHV).

Figure 13 shows the sensitivity of the LCOE to variations in cost and efficiency values for the hydrogen turbine cases. In all cases, off-peak electricity price is the primary cost sensitivity value owing to the low roundtrip efficiency of the system.

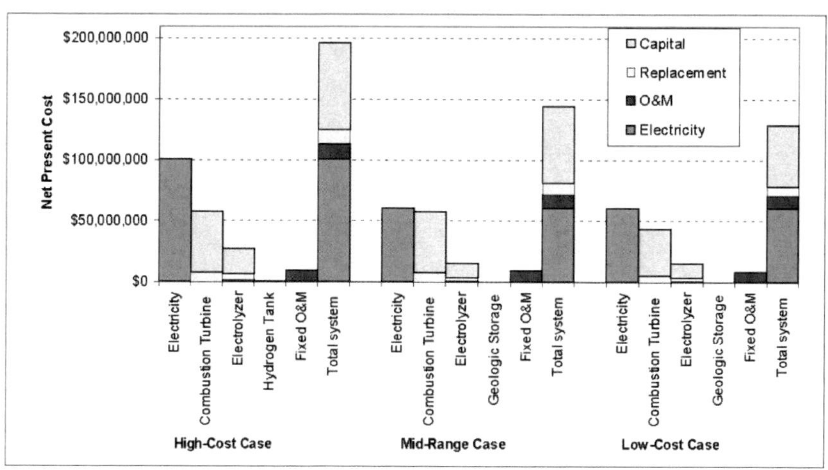

Figure 12. NPC values for the hydrogen turbine cases

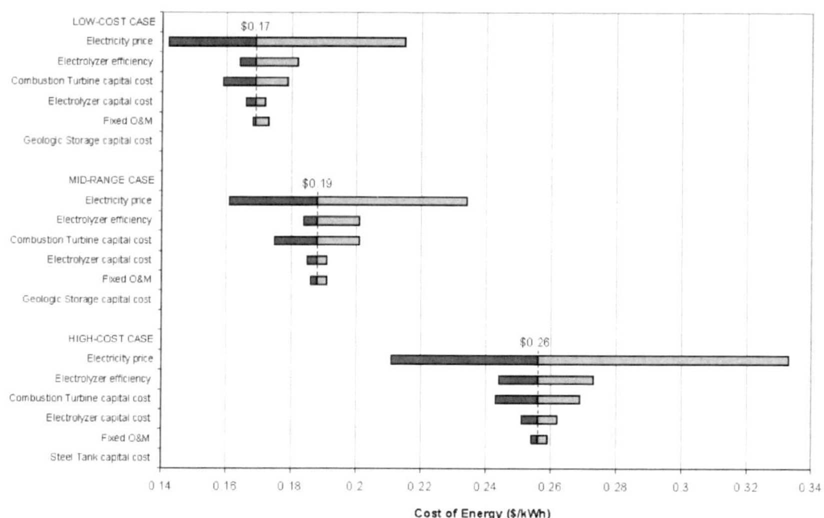

Figure 13. Sensitivity analysis for substitution of a hydrogen expansion combustion or steam turbine for the PEM fuel cell

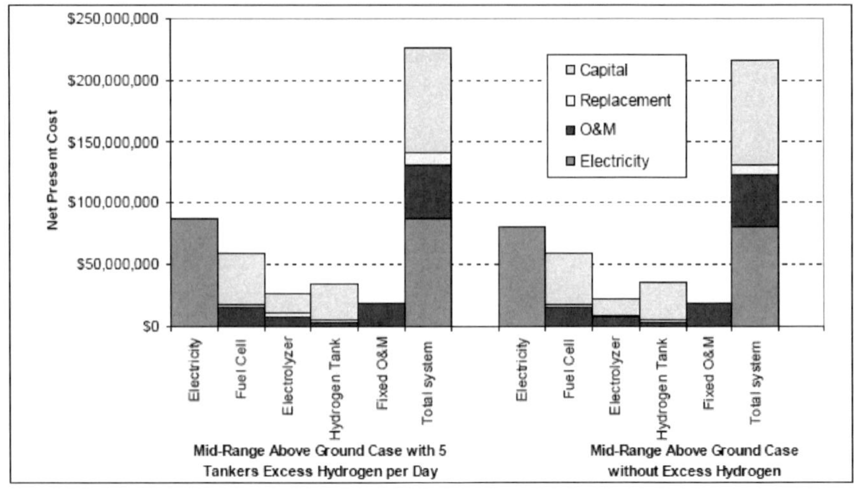

Figure 14. NPC comparison for five tanker trucks of excess hydrogen per day

3.1.5. Excess Hydrogen for the Vehicle Market System Description and Costs

Two scenarios for production of excess hydrogen were evaluated. The mid-range cost cases (aboveground and geologic storage) for the hydrogen energy arbitrage scenario were used as the base case for the analysis of excess hydrogen production.

In the first scenario, it was assumed that demand exists for five gaseous tankers (280 kg per tanker) of excess hydrogen per day. The system was optimized for the hydrogen tank size and electrolyzer system size to meet both the fuel cell demand for hydrogen and supply the gaseous tanker trucks using off-peak electricity (i.e., the electrolyzer is operated for 18 off-peak hours on weekdays and 48 hours over each weekend). It was assumed that the tankers would be filled approximately every two hours. The NPC for the tanker truck excess hydrogen case is presented in Figure 14.

The untaxed levelized cost of hydrogen for this scenario is $4.69/kg. This cost compares to a value of $4.98/kg for hydrogen in the equivalent energy arbitrage scenario. For reference, the current forecourt hydrogen production H2A electrolysis case using the same electricity price ($0.038/kWh) and production level (1,400 kg/day) produces hydrogen at an untaxed levelized cost of $4.00/kg for the production process and $5.05/kg at the dispenser.[9]

A similar comparison was done assuming that hydrogen would be stored in a geologic formation and that excess hydrogen would be fed into a pipeline at the rate of 500 kg/hour (12,000 kg/day). The pipeline demand for hydrogen in this case is approximately equal to the demand for hydrogen for the energy arbitrage scenario. As for the tanker case, it was assumed that the electrolyzer would be operated only during off-peak hours. The NPC for the pipeline excess hydrogen case is presented in Figure 15.

The levelized cost of hydrogen for this scenario is $3.33/kg. This cost compares to a value of $4.21 for hydrogen in the equivalent energy arbitrage scenario. For reference, the current central hydrogen production H2A electrolysis case using the same electricity price ($0.038/kWh) and production level (12,000 kg/day) results in an untaxed hydrogen levelized cost of $6.86.[10]

Sensitivity analyses for the two excess hydrogen cases are presented in Figure 16 and Figure 17. The price of off-peak electricity is the predominant cost driver for the cost of delivered energy in both the excess hydrogen cases. Reducing the price of off-peak electricity from $0.038/kWh to $0.025/kWh reduces the cost of delivered energy by 14% for the five-tanker-per-day excess hydrogen scenario and 13% for the equivalent energy arbitrage scenario (mid-range case with tank storage). The cost of energy is reduced by 19% for the 500 kg/hour case. In contrast, the cost of energy is reduced by 15% for the mid-range geologic storage fuel cell energy arbitrage case. The increased sensitivity of the excess hydrogen cases to the price of off-peak electricity is due to the increased impact of the electrolyzer operating parameters on the overall cost of energy.

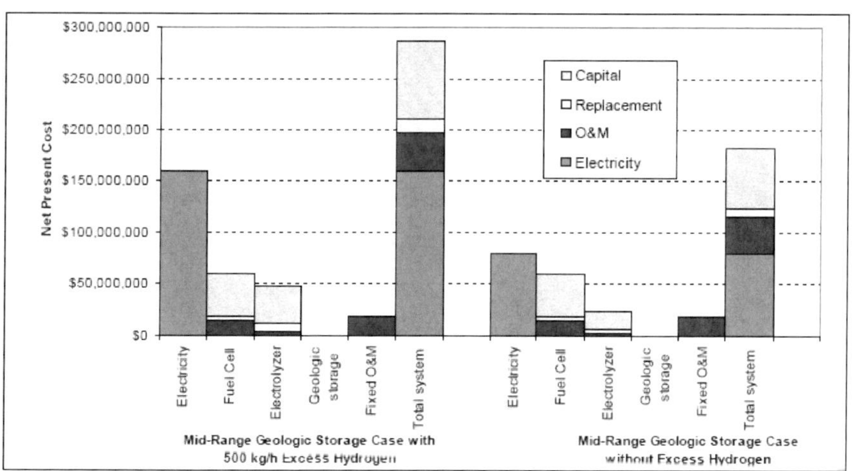

Figure 15. NPC comparison for 500 kg/hour excess hydrogen

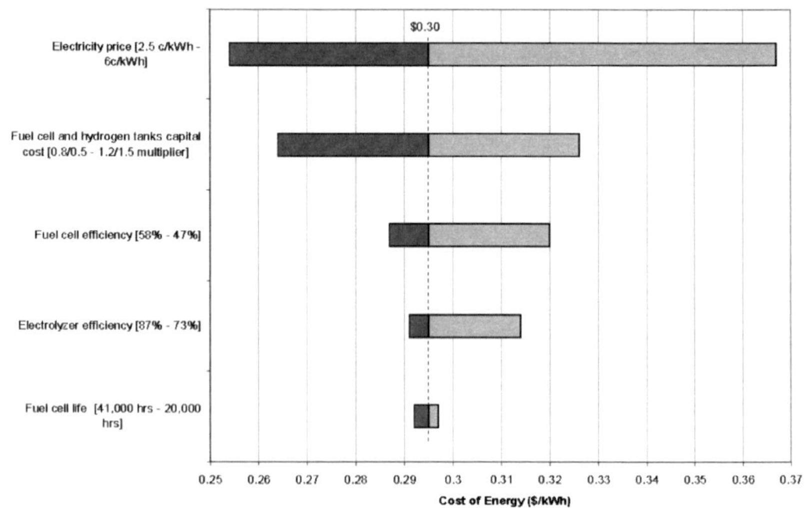

The nominal cost value shown does not line up precisely with the x-axis value due to rounding.

Figure 16. Sensitivity analysis for five tankers per day of excess hydrogen production

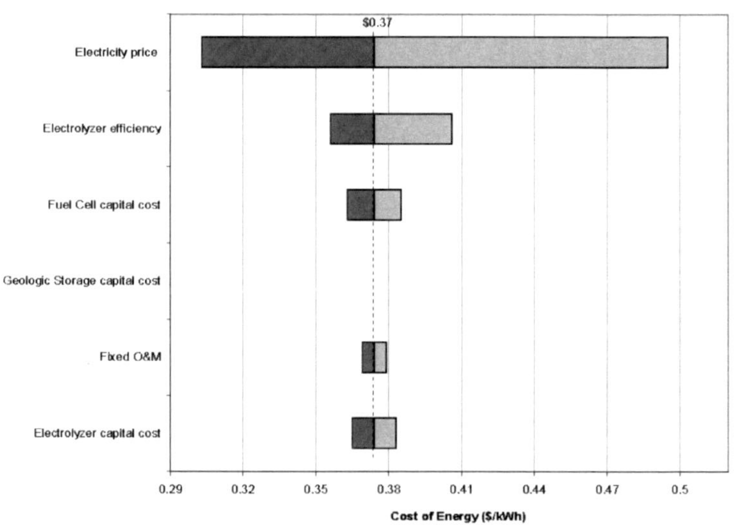

The nominal cost value shown does not line up precisely with the x-axis value due to rounding.

Figure 17. Sensitivity analysis for 500 kg/hour excess hydrogen production

3.2. Batteries

Lead-acid batteries have been used for large-scale energy storage for more than a century, and bulk energy storage systems employing other battery technologies are now being built (Schoenung and Hassenzahl 2003). A nickel-cadmium, 1/2-hour, 26-MW system has been installed in Fairbanks, Alaska. High-temperature sodium sulfur batteries have been installed for bulk energy storage applications in Japan and the United States. Vanadium redox batteries are a newer technology with demonstrated systems up to 1.5 MWh and power ratings up to 1.5 MW.

Costs used in this analysis are primarily derived from two reports by Schoenung (Schoenung and Hassenzahl 2003, Schoenung and Eyer 2008) and two Electric Power Research Institute (EPRI) publications (EPRI-DOE 2003, EPRI 2007). The EPRI reports provide information about the costs included in BoP estimates for batteries. It was assumed that these costs are common to all battery energy storage systems unless otherwise noted. EPRI estimates BoP at $ 100/kW for commercial but not fully integrated systems and $50/kW for fully integrated systems (EPRI-DOE 2003). The EPRI commercial but not fully integrated estimate was used for the mid-range cases. BoP items include the following:

- Project engineering
- Construction management
- Grid connection (transformers, breakers/switches, extension of power lines) connection at 13.8 KV
- Land
- Access
- Procurement
- Permitting

In the EPRI-DOE (2003) report, the battery cost includes an environmentally controlled building, if needed, at $\$100/ft^2 + 20\%$ add-on for a multi-story building. This accounts for roughly $70/kWh or $ 17/kW (EPRI-DOE 2003). For the Schoenung studies, building costs are included in the BoP. BoP costs are presented in $/kWh or $/kW depending on the source of the estimate. Following are descriptions of vanadium redox, nickel cadmium, and sodium sulfur batteries.

Table 9. Literature Values for Vanadium Redox Battery Costs

Source of Estimate	Power Related Costs ($/kW)	Energy Capacity Related Cost ($/kWh)	BoP Cost ($/kW)	Replacement Cost ($/kWh) / Frequency	Fixed O&M ($/kW-yr) / Fixed Costs
Schoenung and Hassenzahl (2003)[1]	175	600	30 ($/kWh)	600/10 yr	20
Schoenung and Eyer (2008)[2]	175	350	30 ($/kWh)	350/10 yr	20
EPRI-DOE (2003)[3]	397	213	100	50% mature price/10 yr	54.8
EPRI (2007) [4] (current 2007$)	1,800	300	500	300 ($/kW)	54.8/250,000
EPRI (2007)[5] (future 2013$)	750	210	500	300 ($/kW)	54.8/280,000
EPRI (2006)	*	300–650[6]	*	*	*
EPRI (2003) [DR/peak shaving 1 MW/4MWh] (prototype/first/ n^{th})	2,260/700/ 500	550/230/1 50	*	*	12/4/2
EPRI (2003) [spinning reserve 10MW/ 20MWh] (prototype/first/ n^{th})	2,150/608/ 426	1,050/410/ 250	*	*	1.2/0.4/0.2
EPRI (2003) [windfarm stabilization/ dispatch 10MW/ 80MWh] (prototype/first/ n^{th})	2,150/608/ 426	300/1 40/ 1 00	*	*	1.2/0.4/0.2

* Not available/not applicable

[1] The application referenced in the study is for a 2.5-MW, 10-MWh facility for Boulder City, UT.

[2] Schoenung and Eyer (2008) use adjusted costs for the battery with respect to the previous study by Schoenung and Hassenzahl (2003). Therefore, replacement and other costs are provided per the previous study.

[3] The application referenced in the study requires that the system provide 10-hour load shifting of stored energy from periods of low value to periods of high value. The reference duty cycle for analysis is scheduled 10 hours, 1 event per day, 250 events per year.

[4] "Capital Cost = $2,300 x (kW rating) + $300 x (kWh rating) + $250,000 (2007 dollars). This equation produces figures within about 5% of the cost figure estimated through the use of the more sophisticated cost model built for this analysis, for systems with power capacities ranging from 200 kW to 10 MW, and with durations from 2 hours to 16 hours."

[5] "Capital Cost = $1,250 x (kW rating) + $210 x (kWh rating) + $280,000 (2013 dollars). This equation produces figures within about 5% of the cost figure estimated through the use of the more sophisticated cost model built for this analysis, for systems with power capacities ranging from 200 kW to 10 MW, and with durations from 2 hours to 16 hours."

[6] VRB Power Systems Inc. was purchased by Prudent Energy in 2009 (www.pdenergy.com). Lower value is projected cost for trial plants 0.25 MW/2 MWh.

3.2.1. Vanadium Redox Batteries

Vanadium redox (reduction-oxidation) batteries, a type of flow battery, are based on the transfer of electrons between forms of vanadium. The power and storage capacity can be specified separately for the particular application. The size of the electrolyte tanks describes the nominal storage capacity, and the size of the cell converter rates the power level capability. Currently, the major suppliers of vanadium redox batteries are Prudent Energy, which purchased VRB Power Systems Inc. in early 2009, and Sumitomo Electric Industries of Japan. There are three somewhat newer companies in the field as well: Cellennium Company, Ltd., V-Fuel Pty Ltd., and Funktionswerkstoffe Forschungs & Entwicklungs GmbH (FWG). The information for vanadium redox batteries has been taken primarily from reports from Sandia National Laboratories (Schoenung and Eyer 2008, Schoenung and Hassenzahl 2003) and EPRI (EPRI 2007, EPRI-DOE 2003). Demonstrated installations range in size from 3 MW for 1.5 seconds of storage to 500 kW for up to 10 hours of storage. Typical power ranges are 100 kW to 3 MW, and storage capacity is typically less than 3 MWh (EPRI-DOE 2003). Cost values for the various systems are presented in Table 9.

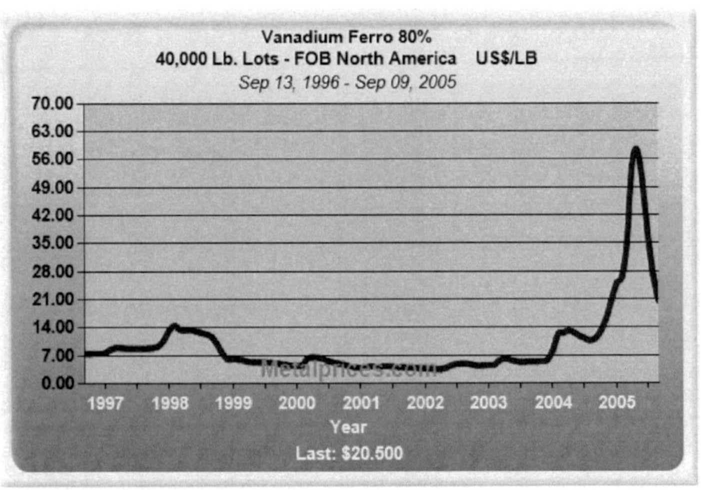

Figure 18. Vanadium prices 1996–2005 (Metalprices.com 2009)

The electrolyte of vanadium redox batteries exhibits little signs of degradation, and the reports suggest that it will not need replacement during the life of the plant. Lifetimes of the electrolyte solution are predicted to be greater than 50 years, at which time the electrolyte can be reused or the more precious vanadium extracted. The capacity of the electrolyte is approximately 20–30 Wh/L (EPRI 2007). If the long-term cost of vanadium pentoxide stabilizes around $4.00–$5.00/lb, it should not become an economic hindrance, but it does constitute approximately 35% of the capital cost of a plant. Prices between 2001 and 2007 have ranged from $1.50–$26.00/lb but are expected to stabilize around $4.00–$5.00/lb (EPRI 2007). Figure 18 illustrates the recent volatility in vanadium prices. The price of the vanadium pentoxide electrolyte for vanadium redox batteries is expected to track vanadium prices.

Capital cost for the electrolyte and electrolyte storage tank are included in energy capacity ($/kWh) costs in Table 9. Power-related costs ($/kW) for vanadium redox batteries include the cell stack and PCS.

3.2.2. Nickel Cadmium Batteries

Nickel cadmium is one of five nickel-based battery types and the most common nickel electrode battery in the utility industry (EPRI-DOE 2003). There are several commercial suppliers of nickel cadmium, including Alcad, Hoppecke Batterien GmbH, Marathon Power Technologies Company, Saft, and Varta. Information regarding the nickel cadmium batteries comes primarily from Schoenung and Eyer (2008), EPRI (2006), Schoenung and Hassenzahl (2003), and EPRI-DOE (2003). The most significant demonstrated utility-scale project is a 13-MWh (26 MW at 30 min.) system outside Fairbanks, Alaska, composed of 13,760 Saft batteries. Nickel cadmium batteries have generally good energy density, more tolerance to abuse, and longer lifetimes than lead- acid batteries; Table 10 lists advantages and disadvantages. Nickel cadmium batteries require enclosure in a temperature-controlled building. Building costs are included in the battery costs for the EPRI studies and in the BoP costs for the Schoenung studies. Cost values for the nickel cadmium batteries are presented in Table 11.

Table 10. Advantages and disadvantages of nickel cadmium batteries

Advantages	Disadvantages

• Durability, long operational life: for sintered plate NiCd life = 3,500 cycles @ 80% depth of discharge = 13 years @ 260 cycles/year[1]
• Lower maintenance cost than lead-acid batteries: sealed batteries require cleaning, vented batteries require adding water several times per year to counter losses from electrolysis and evaporation[2]
  • Visual inspection for damage, leakage, etc.
  • Cleaning
  • Measuring voltage, resistance, specific gravity
  • Water replacement
  • Measuring resistance between terminals
  • Retorquing terminal connections
  • Checking accuracy of DC voltage, DC current, temperature sensors
  • Fixed O&M labor = $50/hr, $900/module per year, includes property taxes and insurance at 2% of initial capital cost
	• High variability with temperature: batteries must be cooled; assume 20% battery-life reduction for every 10°C operating temperature rise
	• Poor charge retention (2%–5% per month at room temperature, vs. lead acid ~1 % per month), highest immediately after charging, higher at higher temperature, mitigated by:
	  • Regular cleaning to avoid deposition of KOH (current path between electrodes)
	  • Charging cold, slight warming before discharge to maximize capacity (highest cold) and discharge voltage (highest at high temperatures)
	• Low power density
	• Gas evolution during operation – H_2 and O_2 evolved from electrolysis of electrolyte during charging; for vented systems, H_2 monitors are needed
	• Higher cost than lead-acid batteries
	• Use of toxic metal cadmium: no exposure during operation, most recycled

[1]Values of 10, 13, and 15 years were used in sensitivity cases.

[2]Vented system assumed.

Table 11. Literature Values for Nickel Cadmium Battery Costs

Source of Estimate	Power Related Costs (PCS) ($/kW)	Energy Capacity Related Costs (Battery) ($/kWh)	BoP ($/kWh)	Replacement ($/kWh) / Frequency	Fixed O&M ($/kW-yr)
Schoenung and Hassenzahl (2003) (bulk storage) ($2003)	125	600[1]	150	600/10 yr	5
Schoenung and Hassenzahl (2003) (Alaska plant) ($2003)[2]	250	1,360[3]	150	*	5
Schoenung and Hassenzahl (2003) (distributed generation) ($2003)	175	600	50	600/10 yr	25
Schoenung and Eyer (2008) (bulk storage) ($2003)	125	600	150	600/10 yr	5
EPRI-DOE (2003)[4] ($2003)	150	1,197	$1 00/kW	Variable[5]	26.5
EPRI-DOE (2003) range[6] ($2003)	144–153	1,197–1,424	$100/kW	Variable[5]	15.1–26.5
EPRI (2006)	*	500–600[7]	*	*	*

* Not available/not applicable

[1] Manufacturer predicted cost. Demonstrated plant was ~$900/kWh.

[2] Schoenung and Hassenzahl (2003) actual costs for 13 MWh (1/2 hour, 26 MW) plant near Fairbanks Alaska.

[3] Schoenung and Hassenzahl (2003); $900/kWh for batteries + $460/kWh for other facilities (e.g., transmission lines) not included in BoP.

[4] The application referenced requires that the system continuously detect and mitigate infrequent SPQ events lasting to 5 seconds. In addition, the system will provide load-shifting services at 1.8 MWac for 3 hours per day for 60 days per year. The regular deep-cycling of this application requires use of a sintered-plate nickel cadmium cell. The lifetime of this system would be affected by both calendar life and cycle life; the system can be expected to last 10 years. Disposal costs of NPV $1 .4/kW would also be incurred.

[5] Nickel cadmium lifetimes are dependent on depth of discharge and number of cycles, typically in the 10–15 year range for light-cycling applications. The replacement cost will go along the lines of the original cost.

[6] Systems from 2.5 to 5 MWac.

[7] "NiCd estimated OEM battery costs."

3.2.3. Sodium Sulfur Batteries

Currently, the only supplier of sodium sulfur batteries is NGK Insulators, Ltd. of Japan. Originally developed for electric vehicles, more recently the Tokyo Electric Power Company (TEPCO) has developed several utility-scale projects with NGK. The demonstration projects range from 500 kW to 6 MW in scale, including two 48-MWh plants. Information has been collected from Schoenung and Hassenzahl (2003) and EPRIDOE (2003). Cost values for sodium sulfur batteries are presented in Table 12.

3.2.4. Net Present Costs and Sensitivity Analyses for Battery Systems

The HOMER model treats a variety of batteries, including a vanadium redox flow battery, and allows the user to define the characteristics for new batteries. In general, the batteries included in HOMER are for small-scale distributed energy storage applications. New batteries were defined for the nickel cadmium and sodium sulfur batteries included in this analysis. The characteristics of these batteries were developed based on the values presented in EPRI-DOE (2003). The vanadium redox battery included in HOMER was modified for the analysis. Batteries are assumed to have DC input and output. Therefore, an AC/DC converter was added to the HOMER system configuration. The inverter and rectifier efficiencies were assumed to be 95% (or ~90% roundtrip) for all the battery cases. Costs for the converter are derived from the PCS costs given in the literature. Table 13 presents the high-cost, mid-range, and low-cost values used in HOMER for the battery systems. All the costs were developed for the six-hour, 50-MW energy storage scenario.

Net present costs for the nickel cadmium battery systems are presented in Figure 19. Both the cost for electricity to charge the batteries and the battery initial capital costs decrease because of the increased efficiency of the batteries for the lower-cost cases. For the high- cost, mid-range, and low-cost cases, the battery DC roundtrip efficiencies are 60%, 65%, and 70%, respectively (EPRI-DOE 2003). The AC-to-AC roundtrip efficiencies are 54%, 59%, and 63%, respectively. Replacement costs are higher for the mid-range case because of differing assumptions in the reference studies.

Table 12. Literature Values for Sodium Sulfur Battery Costs

Source of Estimate	Power Related Cost (PCS) ($/kW)	Energy Capacity Related Cost (Battery) ($/kWh)	BoP ($/kWh)	Replacement ($/kWh)	Fixed O&M($/kW-yr)
Schoenung and Hassenzahl (2003) (bulk storage)	**150**	**250**	**50**	**230 (10-yr life)**[1]	20
Schoenung and Hassenzahl (2003) (distributed generation)	**150**	**250**	**0**	**230 (10-yr life)**[1]	20
EPRI-DOE (2003)[2]	**204**	**196**	**$100/kW**	**Variable**[3]	51.2
EPRI-DOE (2003) range[4]	**202–289**	**196–508**	**$100/kW**	**Variable**[3]	19.2–51.2
EPRI (2006)	*	250–400[5]	*	*	*

* Not available/not applicable

[1] Estimate based on 250 six-hour cycles per year.

[2] This application requires that the system provide 10-hour load shifting, regulation control, and spinning reserve functions on a scheduled basis using a Type II PCS (i.e., prompt PCS response is not required and no PCS standby losses occur). Two hundred fifty-eight (258) NAS E50 Modules capable of discharging at a pulse factor of 1 (i.e., 50 kW per module) for up to 8.6 hours equipped with a programmable PCS will provide load shifting for 10 hours per day at 10 MWac for 250 days per year. The projected battery life for this application is 15 years because cycle life (as measured by the cumulative cycle fraction of 89% at 90% DOD) exceeds shelf life.

[3] Replacement for applications of 20 years based on mature prices.

NAS Module	2006 Prices, K$	Mature Prices, K$
E50	$75	$55
G50	$68	$50
PQ50	$75	$55

[4] Range from 10 to 100 MWac.

[5] NGK Web site projected cost for trial plants 6 MW/48 MWh and 1 MW/8 MWh (EPRI 2006).

Table 13. HOMER Model Cost Input Values for Battery Energy Storage Systems ($2008)

	Energy Capacity Related Cost (Battery) ($/kWh)	Power Related Cost (PCS) ($/kW)	BoP ($/kWh)	Fixed O&M($/kW-y)
Nickel Cadmium				
High-CostCase[1]	1,570	2882	173	5.8
Mid-Range Case[2]	1,380	1503	115 ($/kW)	31
Low-CostCase[4]	690	144	173	5.8
Sodium Sulfur				
High-CostCase[5]	288	173	58	23
Mid-Range Case[6]	226	235	115 ($/kW)	59
Low-Cost Case	30% reduction from mid-range case[2]	173	58	59
Vanadium Redox[7]				
High-CostCase[8]	300	1,800	500 ($/kW)	54.8
Mid-Range Case[9]	210	750	500 ($/kW)	54.8
Low-Cost Case[9]	210	30% reduction from mid-range case	500 ($/kW)	54.8

[1] Schoenung and Hassenzahl (2003). Actual costs for Fairbanks Alaska facility.

[2] EPRI-DOE (2003).

[3] PCS cost is derived from equation in EPRI-DOE (2003) for a programmed response PCS without VAR support; $/kW ($2003) = 11,500 * Vmin-0.59 where Vmin is the minimum discharge voltage (maximum current).

[4] Schoenung and Eyer (2008).

[5] Schoenung and Hassenzahl (2003), Schoenung and Eyer (2008). Replacement costs at $230/kWh.

[6] Values from EPRI-DOE (2003), NKG Insulators Ltd, E50 peak shaving battery (50-kW modules).

[7] Electrolyte costs are not expected to decrease in the future due to the cost of vanadium. Electrolyte makes up about 30% of the capital cost of the system. However, future improvements in the system are expected to result in some cost reduction. Electrolyte costs decrease from $256/kWh to $151/kWh for the future case.

[8] EPRI (2007) "present day" costs. Replacement cost for cell stack only at "future" cost.

[9] EPRI (2007) "future" costs. Replacement cost for cell stack only at "future" cost.

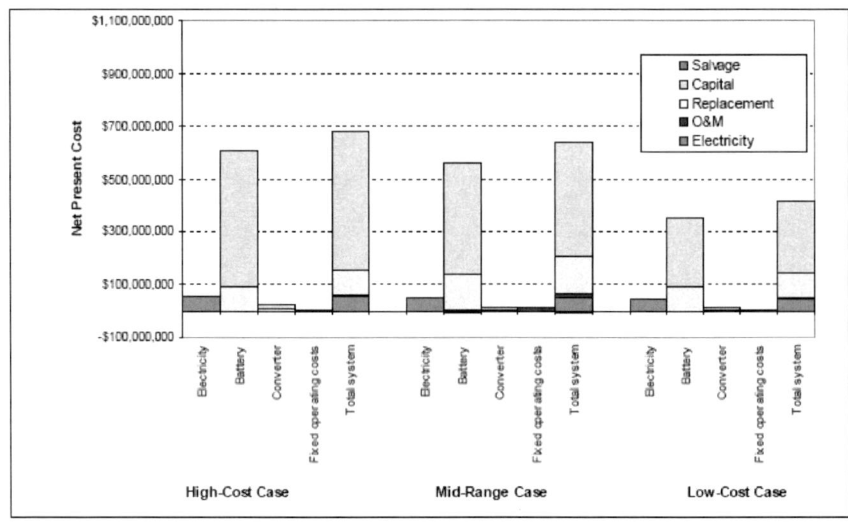

Figure 19. NPC for nickel cadmium battery systems

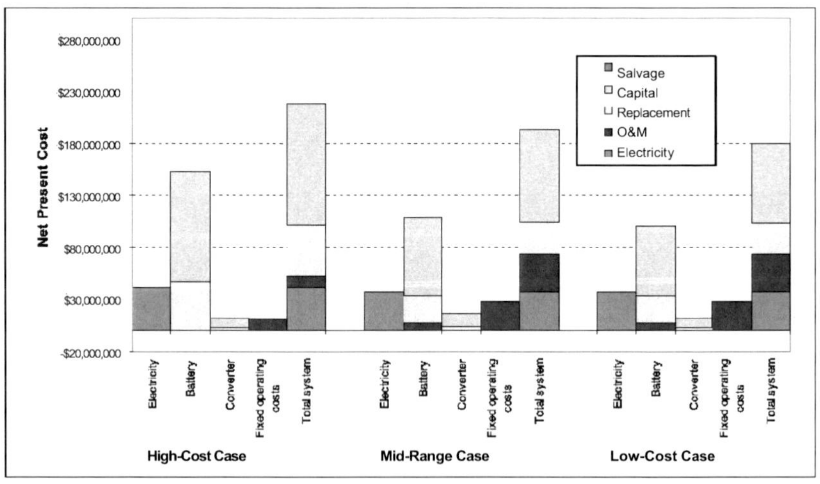

Figure 20. NPC for sodium sulfur battery systems

Figure 20 presents the NPC for the sodium sulfur battery systems. For the high-cost case, the battery DC roundtrip efficiency is 78% (Schoenung and Hassenzahl 2003, Schoenung and Eyer 2008). The battery DC roundtrip efficiency for the mid-range and low-cost cases is 85% (EPRI-DOE 2003).The system AC-to-AC roundtrip efficiencies are 71% for the high-cost case and 77% for the mid-range and low-cost cases. Replacement costs are higher for the high-cost case because of differing assumptions in the reference studies.

Figure 21 presents the NPC for the vanadium redox battery systems. The DC roundtrip efficiency is assumed to be 80% for all cases (EPRI 2007). The AC-to-AC efficiency is 72%. Capital costs for the vanadium redox battery system are a higher percentage of the total cost than for the sodium sulfur battery system. However, because the electrolyte makes up a large fraction of the initial capital cost and does not have to be replaced, the replacement costs for the vanadium redox system are slightly lower than for the sodium sulfur battery system and are a smaller percentage of the overall cost.

Figure 22, Figure 23, and Figure 24 show the sensitivities of the various battery systems to variation in capital and operating costs. In most cases, the systems were most sensitive to uncertainty in capital costs with the price of off-peak electricity as the second most influential factor.

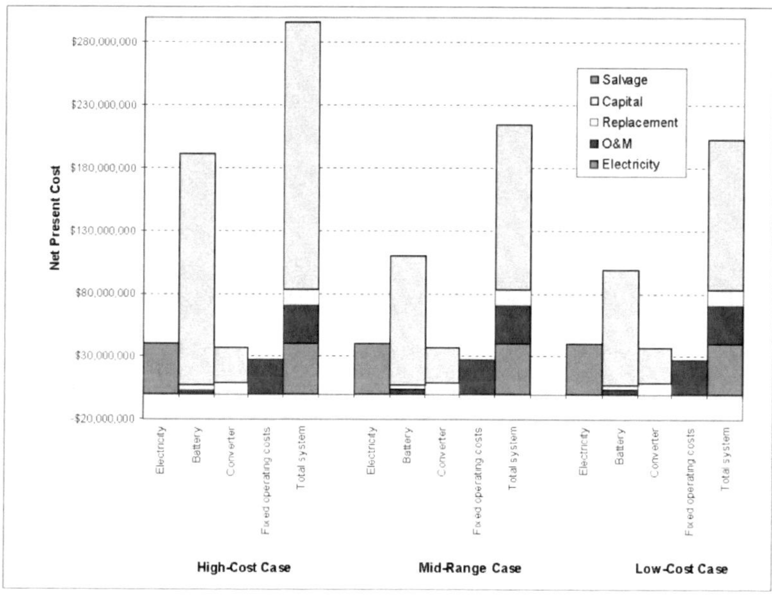

Figure 21. NPC for vanadium redox battery systems

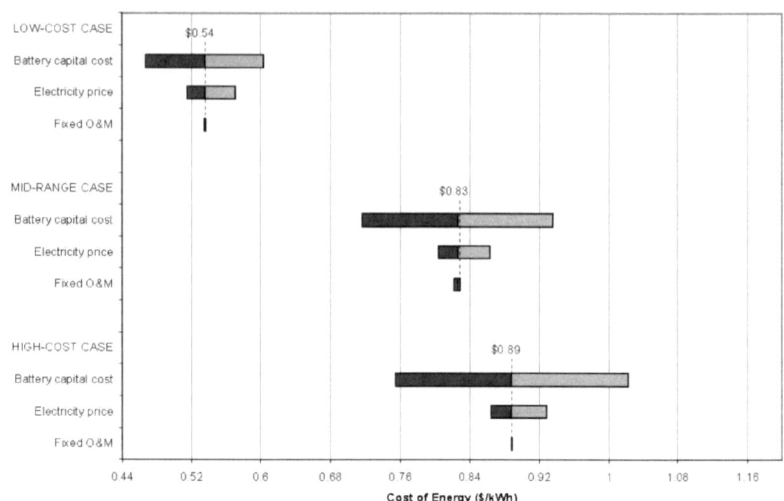

Figure 22. Sensitivity analysis for nickel cadmium battery systems.

3.3. Pumped Hydro

Pumped hydro is a mature technology. The first plant built in the United States in 1928– 1929 featured two 3-MW reversible turbines. Today, U.S. pumped hydro capacity is about 19,000 MW (Schoenung and Hassenzahl 2003). Costs, drawn from Schoenung and Hassenzahl (2003), are presented in Table 14.

Table 14. Literature Values for Pumped Hydro Costs

Source of Estimate	Power Related Cost (Reversible Turbine) ($/kW)	Energy Capacity Related Cost (Storage System) ($/kWh)	BoP ($/kWh)	Fixed O&M ($/kW-yr)
Schoenung and Hassenzahl (2003)	1,000(constant speed) / 1,050 (variable speed)	10	4	2.5

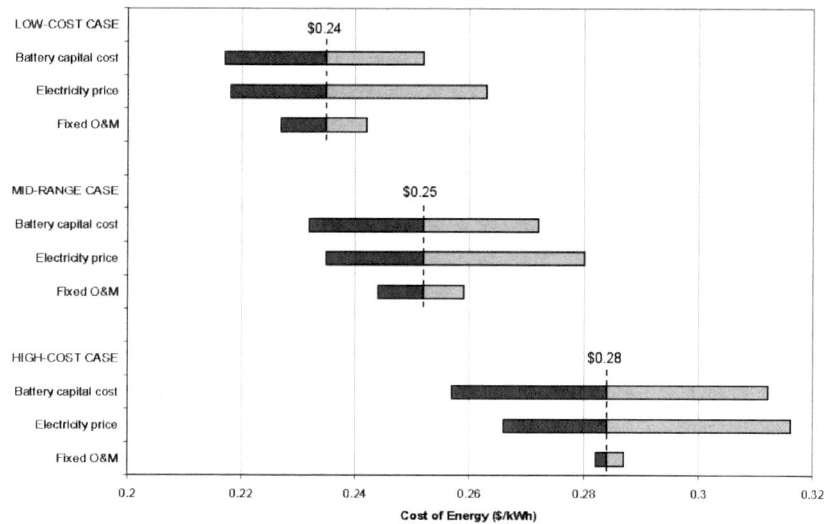

Figure 23. Sensitivity analysis for sodium sulfur battery systems

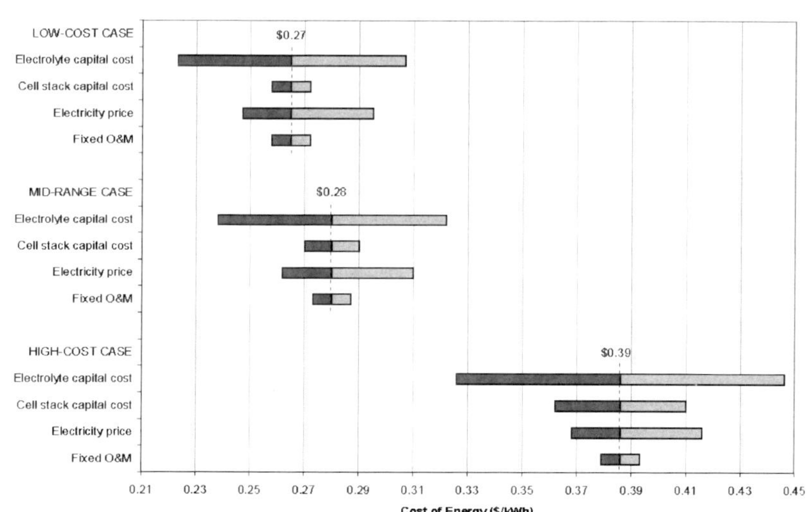

Figure 24. Sensitivity analysis for vanadium redox battery systems

Table 15. Capacity and Cost Information for 1,000-MW and Larger Pumped Hydro Installations Worldwide (Electricity Storage Association 2009)

Location	Plant Name	On-Line Date	Hydraulic Head (m)	Max Total Rating (MW)	Hours of Discharge	Plant Cost
Australia	Tumut 3	1973		1690		
China	Tianhuangping	2001	590	1800		$1080 M
	Guangzhu	2000	554	2400		
France	Grand Maison	1987	955	1800		
Germany	Markersbach	1981		1050		
	Goldisthal	2002		1060		$ 700 M
Iran	Siah Bisheh	1996		1140		
Italy	Piastra Edolo	1982	1260	1020		
	Chiotas	1981	1070	1184		
	Presenzano	1992		1000		
	Lago Delio	1971		1040		
Japan	Imaichi	1991	524	1050	7.2	
	Okuyoshino	1978	505	1240		
	Kazunogowa	2001	714	1600	8.2	$3200 M
	Matanogawa	1999	489	1200		
	Ohkawachi	1995	411	1280	6	
	Okukiyotsu	1982	470	1040		
	Okumino	1995	485	1036		
	Okutataragi	1998	387	1240		
	Shimogo	1991	387	1040		
	Shin Takesagawa	1981	229	1280	7	
	Shin Toyne	1973	203	1150		
	Tamahara	1986	518	1200	13	
Luxembourg	Vianden	1964	287	1096		
Russia	Zagorsk	1994	539	1200		
	Kaishador	1993		1600		
	Dneister	1996		2268		
South Africa	Drakensbergs	1983	473	1200		
Taiwan	Minghu	1985	310	1008		$ 866 M
	Mingtan	1994	380	1620		$ 1338 M
U.K./Wales	Dinorwig	1984	545	1890	5	$ 310 M
USA / CA	Castaic	1978	350	1566	10	
USA / CA	Helms	1984	520	1212	153	$ 416 M
USA / MA	Northfield Mt	1973	240	1080	10	$ 685 M
USA / MI	Ludington	1973	110	1980	9	$ 327 M
USA / NY	Blenheim-Gilboa	1973	340	1200	12	$ 212 M
USA / NY	Lewiston (Niagara)	1961	33	2880	20	
USA / SC	Bad Creek	1991	370	1065	24	$ 652 M
USA / TN	Racoon Mt	1979	310	1900	21	$ 288 M
USA / VA	Bath County	1985	380	2700	11	$1650 M

The Electricity Storage Association has collected information on existing pumped hydro facilities that are larger than 1,000 MW in power capacity. Table 15, reproduced from the Electricity Storage Association Web site, lists information for the facilities, including total capital costs for several. The most recent U.S. plant is the Bad Creek South Carolina pumped hydro plant, which was operational in 1991. The total cost per kW for that plant, escalated to 2008 dollars, was used for the low-cost case for the analysis. The value for fixed operating costs and roundtrip efficiency were taken from Schoenung and Hassenzahl (2003) for a variable-speed turbine.

Table 16. Pumped Hydro Storage System Costs ($2008)

	Storage System Including PCS	BoP ($/kWh)	Fixed O&M ($/kW-y)
High-cost case	$12/kWh + $1,209/kW	5	2.9
Mid-range case	$12/kWh + $1,151/kW	5	2.9
Low-cost case	$1 2/kWh + $888/kW	0	2.9

Pumped hydro requires terrain with enough elevation change to allow for power generation from potential energy. Schoenung and Hassenzahl (2003) showed that a large volume of water is required. Approximately 400 m^3 of water is required per meter of reservoir height per kWh generated; for example, for a reservoir elevated 100 m above the turbine, 4 m^3 of water are required for each kWh generated. The environmental impact associated with developing reservoirs large enough to generate power may preclude new projects.

3.3.1. Net Present Costs and Sensitivity Analyses for Pumped Hydro Systems

Values used in the analysis for pumped hydro system costs are shown in Table 16. The pumped hydro turbine is assumed to last the entire life of the storage system (Schoenung and Hassenzahl 2003).

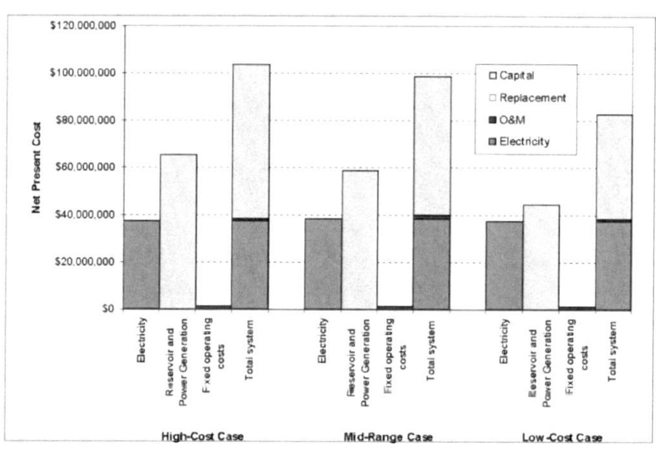

Figure 25. NPC for pumped hydro systems

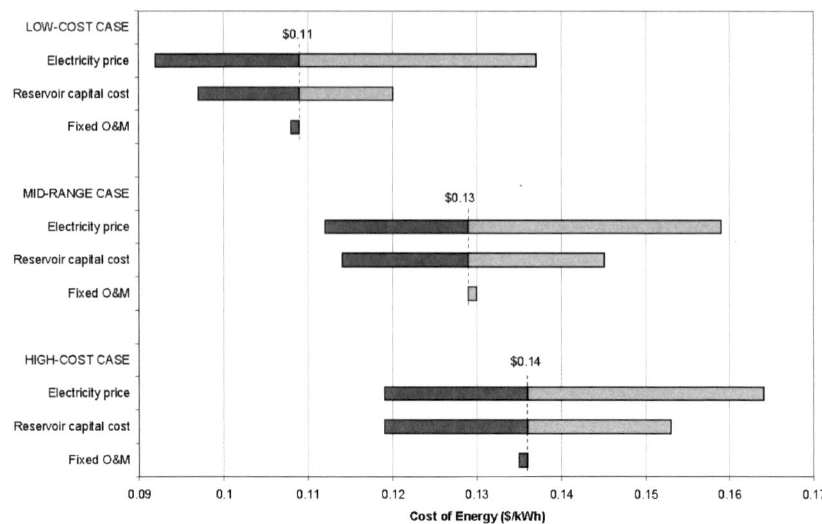

Figure 26. Pumped hydro LCOE sensitivity analysis.

Net present cost values and sensitivity analysis results are presented in Figure 25 and Figure 26. For pumped hydro systems, there are essentially no replacement costs. Routine maintenance of the turbines and reservoir are included in fixed operating and maintenance (O&M) costs. O&M costs for pumped hydro are on the order of 1/10th the fixed O&M costs for the mid-range fuel cell systems and 1/10th to 1/20th of the cost for battery systems. Because of the high roundtrip efficiency of pumped hydro systems (75% to 78%), capital costs are a more significant cost driver than for the hydrogen systems.

3.4. Compressed Air Energy Storage

Compressed air energy storage for large applications uses a compressor during off-peak times to compress air into underground caverns, generally salt mines, aquifers, etc., and then run a high-temperature combustion turbine during peak hours. This makes the combustion turbine more efficient during peak hours because it is not running a compressor as well. In a typical combustion turbine system, 55% to 70% of the expander power is used to drive the compressor (EPRI-DOE 2003).

Natural gas and oil companies have used underground storage for 80 years, and approximately 80% of the United States has suitable geography for gas storage. Multiple CAES concepts have been developed to cover a range of utility-scale needs (EPRI-DOE 2003), but there are only two major existing CAES installations: in Huntorf, Germany, built in the 1970s, and McIntosh, Alabama, built in the 1990s. Plants, built and proposed, range in size from 110 to 2,700 MW. For distributed applications, underground storage might be replaced with steel tank storage options. However, there are no demonstrated surface CAES plants.

The primary components of the CAES system are a compressor with intercooling to compress the air to a pressure of approximately 1,000 psi (EPRI-DOE 2003) for storage in a geologic formation 1,500 to 2,500 feet below the surface. Pressure within the storage cavern drops to about 650 psi during discharge. During the discharge cycle, air extracted from the cavern is either heated by exhaust from the combustion turbine in a recuperator or directly fed with fuel (typically natural gas) into a combination of high- and low- pressure expanders and combustors. The recuperator reduces fuel consumption in the expander/generator train by about 25%. The heat rate, defined as Btus fuel used per kWh net electricity output, is about 4,100 Btu/kWh for the McIntosh plant. Because additional energy is derived from the fuel, a CAES plant provides 25% to 60% more electricity during the power-generation cycle than is used for compression of the air during off-peak hours (EPRI-DOE 2003). Figure 27 presents a schematic of a CAES plant with a recuperator similar to the design of the McIntosh plant (Greenblatt et al. 2004).

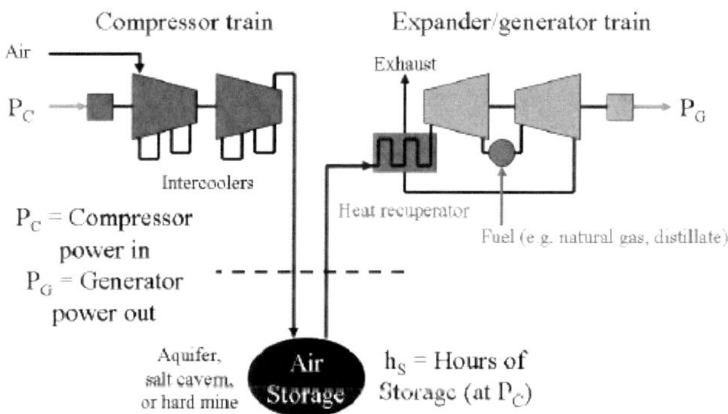

Figure 27. Schematic of CAES plant including heat recuperation (Greenblatt et al. 2004).

Table 17. Literature Values for CAES Costs

Source of Estimate	Power Related Cost (PCS) ($/kW)	Energy Capacity Related Cost (Storage System) ($/kWh)	BoP ($/kWh)	Fixed O&M ($/kW-yr)
Schoenung and Hassenzahl (2003) (bulk storage)	425	3	50	2.5
Schoenung and Hassenzahl (2003) (distributed generation/surface)	550	120	50	10
Schoenung and Eyer (2008) (distributed generation/surface)	550	120	50	*
EPRI-DOE (2003) (range 2002$)[1]	400–450	*	*	*
EPRI-DOE (2003) (salt mine 300 MWac)	270	1[2]	$170/kW	13
EPRI-DOE (2003) (surface 10 MWac)	270	40	$160/kW	19–24.6
Van der Linden (2006)	500–600[3]	*	*	*
EPRI (2003) (salt/ porous/hard rock/surface)	350 (all)	1/0. 1 0/30/30	*	6
EPRI-DOE (2004) (salt/surface)[4]	300	1 .75[5]/40	$ 210/ $200/kW	(23.6–24.6)/ (27–32.6)
Nakhamkin (2007) (72 MW adiabatic CAES)	1,700	*	*	6[6]

* Not available/not applicable

[1] Plants: 290 MW/10 hr, 110 MW/26 hr, 2,700 MW/30 hr (never completed), 540 MW/NA (canceled).

[2] “The reference energy storage capacity for large CAES technologies is 10 hours. A representative price for CAES systems over the range of 8 to 20 hours of storage can be obtained by applying increments/decrements at the rate of $1/kWh.”

[3] 100–300 MW.

[4] This is an update to the EPRI-DOE (2003) handbook.

[5] "The reference energy storage capacity for large CAES technologies is 10 hours. A representative price for CAES systems over the range of 8 to 40 hours storage can be obtained by applying increments/decrements at the rate of $1 .75/kWh."

[6] Value from EPRI (2003).

The Alabama plant was designed and constructed in three years; smaller plants might be completed within a year (EPRI-DOE 2003).

Cost information for CAES systems, presented in Table 17, was collected from several sources:

- Schoenung and Eyer (2008)
- Nakhamkin (2007)
- van der Linden (2006)
- EPRI-DOE (2004)
- Schoenung and Hassenzahl (2003)
- EPRI-DOE (2003)
- EPRI (2003)

3.4.1. Net Present Costs and Sensitivity Analyses for Compressed Air Energy Storage Systems

Values used in the analysis for CAES system costs are shown in Table 18. A simple CAES plant design without heat recuperation was assumed for the high-cost case. A heat rate of 6,000 Btu/kWh was assumed for this plant. A heat rate of 4,000 Btu/kWh, similar to the McIntosh plant, was assumed for the mid-range case, and a value of 3,800 Btu/kWh, for an advanced design, was assumed for the low-cost case (Nakhamkin et al. 2007). An energy ratio of 0.7 was assumed in all cases. The compressor and combustion turbine systems were assumed to have a 10-year life. The compressor was assumed to have an efficiency of 95%.

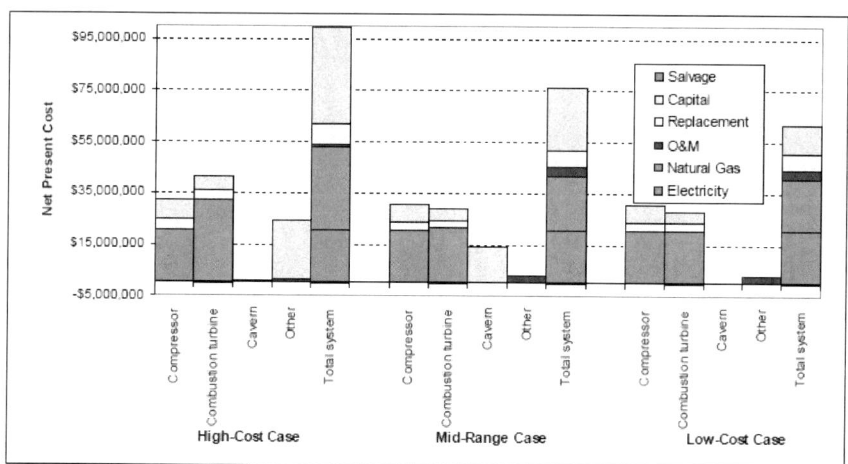

Figure 28. NPC for CAES systems

Net present cost values and sensitivity analysis results are presented in Figure 28 and Figure 29. For the CAES system, the compressed air provides 70% of the output energy while the natural gas supplies the remainder. In this respect, CAES is a hybrid between an energy storage system and a generator. The large proportion of NPC associated with fuel (55%) results from this dual function. Off-peak electricity price and natural gas price are the primary cost drivers for sensitivity analysis. In the mid-range case, it was assumed that the cavern would be mined from hard rock (see Table 3 and Table 17), resulting in the high cavern development cost sensitivity for that scenario. For the purpose of this study, AC-to-AC roundtrip efficiency for the CAES system is defined as the total electricity output divided by the total energy input (electricity plus natural gas). The ACto-AC roundtrip efficiencies for the three scenarios are 41%, 53%, and 55% for the high- cost, mid-range, and low-cost cases, respectively.

An additional scenario was evaluated using cost and efficiency values for an adiabatic CAES system proposed by Nakhamkin (2007). This system has a roundtrip efficiency of 75%, which is higher than the other CAES systems analyzed. However, the high capital cost associated with addition of the heat storage system resulted in a higher NPC: $140 million compared with $99 million for the high-cost case. The LCOE was $0.1 8/kWh for the adiabatic CAES system compared with $0.1 3/kWh for the high-cost case.

Table 18. CAES Storage System Costs ($2008)

	Storage System Including PCS	**BoP($/kWh)**	**Fixed O&M ($/kW-y)**	**Natural Gas HeatRate (Btu/kWh)**
High-cost case	$3.45/kWh + $490/kW	58	2.9	6,000
Mid-range case	$34.54/kWh + $403/kW	0	6.9	4,000
Low-cost case	$1.15/kWh + $403/kW	0	6.9	3,800

4. CONCLUSIONS

The results of this study enable comparison of the cost of hydrogen and several competing technologies for energy storage, including the cost of producing excess hydrogen for use in vehicles. Each technology also has various non-economic benefits and drawbacks.

4.1. Energy Arbitrage Benchmarking Cost Analysis

Hydrogen for energy storage is potentially cost competitive with battery systems but not competitive with pumped hydro or CAES systems for the scenarios evaluated here. Figure 30 summarizes the comparison of levelized (annualized total capital and operating) cost of delivered electricity for hydrogen (green bars) and competing technologies (blue bars). The bottom of the bars represents the low end of the range for the low-cost cases, and the top of the bars represents the high end of the range for the high-cost cases. The numerals shown are the nominal values of the mid-range cases; these mid-range values do not represent a statistical determination of most-probable costs.

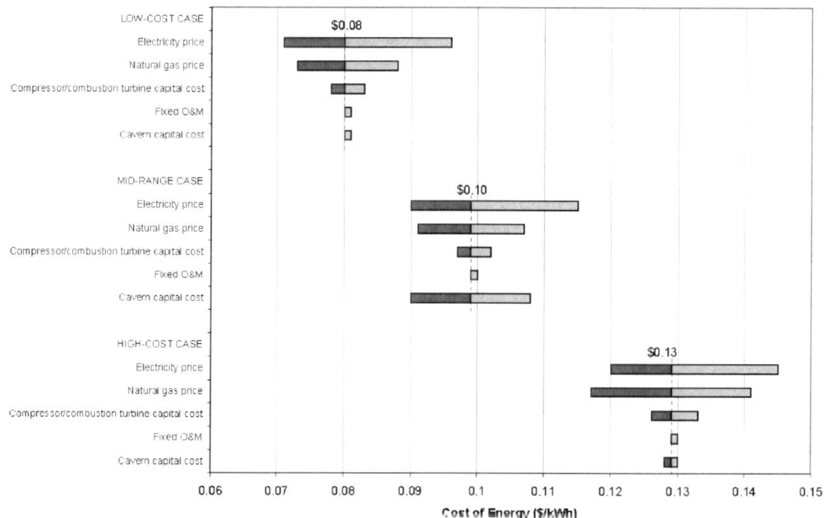

Figure 29. Sensitivity analysis for CAES systems

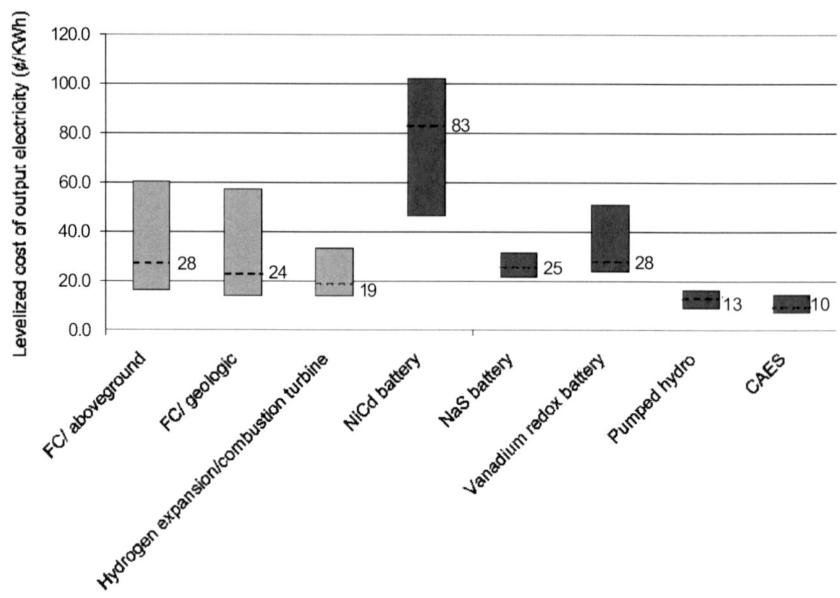

Figure 30. Ranges of LCOE for electricity storage systems

The cost range for each system reflects the cost ranges found in the literature and estimates of potential cost reductions as technologies develop. The fuel cell scenario cost range reflects the comparative immaturity of fuel cell technologies for this application. It is anticipated that costs for fuel cells will decrease as the technology matures and is implemented in more applications. Hydrogen combustion turbines could prove to be viable for energy storage applications and have the potential for providing additional flexibility to utilities through co-firing of mixtures of natural gas and hydrogen. The fuel cell systems have a relatively large range: from $0.18/kWh to $0.50/kWh for the low and high base cases (without sensitivities), respectively. The difference is primarily due to the potential for significant cost reductions for fuel cells in the near future. The battery systems are expected to decrease in cost as the technologies become better established. However, it is unlikely that a nickel cadmium system would be economical for the scenario evaluated here.

Table 19. Roundtrip (AC-to-AC) Efficiencies for Storage Systems in This Study

System (Mid-Range Case @ $0.038/kWh)	Roundtrip Efficiency (%)
Fuel cell/aboveground storage	34 (LHV)
Fuel cell/geologic storage	35 (LHV)
Hydrogen expansion/combustion turbine	48 (LHV)
CAES[1]	53
Nickel cadmium battery	59
Sodium sulfur battery	77
Vanadium redox battery	72
Pumped hydro	75

[1]AC-to-AC roundtrip efficiency for the CAES system is defined as the total electricity output divided by the total energy input (electricity plus natural gas).

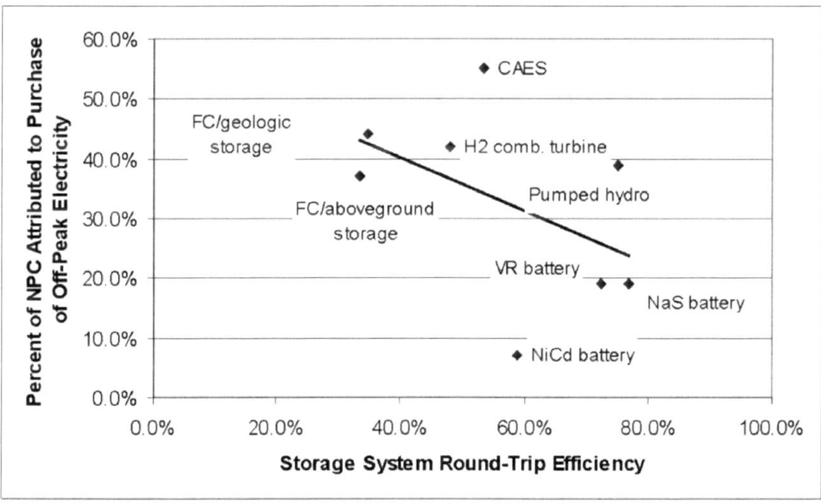

Figure 31. Off-peak electricity percentage of NPC versus system roundtrip efficiency

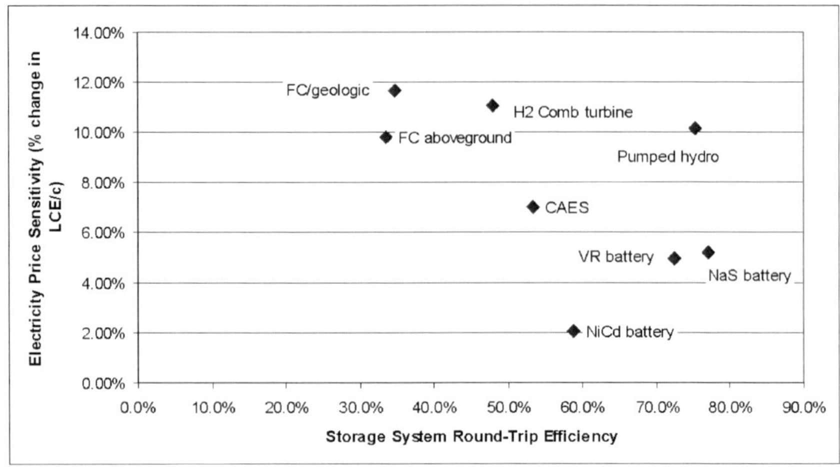

Figure 32. Electricity price sensitivity (% change in LCOE per $0.01 change in off-peak electricity price)

The roundtrip AC-to-AC efficiency varies between the technologies evaluated and has an effect on the economics for each system. Systems with lower roundtrip efficiency require larger storage volumes to accommodate the increased fuel requirements for reconversion back into electricity and larger equipment (more fuel cells for example) to meet the same requirement for output electricity. Low roundtrip efficiency also increases the amount paid for purchase of off-peak electricity over the lifetime of the facility. Figure 31 plots the percentage of NPC attributed to purchase of off-peak electricity (and natural gas in the CAES system) and the storage system roundtrip efficiency. All efficiencies for the hydrogen systems are presented as LHV. The solid line on the figure is a linear fit of the data and indicates that, in general, the proportion of the facility's NPC resulting from purchase of off-peak electricity decreases with increasing system efficiency. Table 19 lists the roundtrip efficiencies for the storage systems analyzed. The efficiency values presented are those used for the mid-range case for all technologies.

The storage system's sensitivity to the cost of input electricity is also influenced by the system roundtrip efficiency. Figure 32 plots the electricity price sensitivity (% change in LCOE per $0.01 change in off-peak electricity price) and storage system efficiencies.

Table 20. Non-Economic Benefits and Drawbacks of Energy Storage Technologies

	System Operation		Environmental	
	Benefits	**Drawbacks**	**Benefits**	**Drawbacks**
Hydrogen	• Modular (can size electrolyzer separately from FC or turbine to produce extra hydrogen) • Very high energy density for compressed hydrogen (~100 times the energy density for compressed air at 120 bar ΔP, IGCC gas turbine) • System can be fully discharged at all current levels	• Low electrolysis/FC round trip (AC- to-AC) efficiency (34%–35%) • Low roundtrip efficiency when hydrogen used in a combustion turbine (~48%) • Hydrogen storage in geologic formations other than salt caverns may not be feasible • Electrolyzers and fuel cells require cooling	• Catalyst can be reclaimed at end of life	• Environmental impacts of mining and manufacturing of catalyst • Low roundtrip efficiency increases emissions impact of energy conversion losses
Batteries	• Modular system allows replacement of individual batteries and addition of storage capacity • Mid-range to high roundtrip efficiency (65%–75%)	• Voltage-to-current relationship limits amount of energy that can be extracted, especially at high current • Some battery systems require climate controlled building	• High roundtrip efficiency reduces emissions impact of energy conversion losses	• Toxic and hazardous materials

Table 20. (Continued)

	System Operation		Environmental	
	Benefits	**Drawbacks**	**Benefits**	**Drawbacks**
Pumped Hydro	• Well established and simple technology • High roundtrip efficiency (70%– 80%)	• Large water reservoir/suitable reservoir location required • Requires mountainous terrain or other method for producing potential energy • Water loss due to evaporation and seepage ~6.9 gal/kWh for conventional hydroelectric power • Extremely low energy density (0.7 kWh/m^3)	• No toxic or hazardous materials	• Large water losses due to evaporation, especially in dry climates • Habitat loss due to reservoir flooding • Stream flow and fish migration disruption if associated with conventional hydroelectric power generation
CAES	• Proposed advanced designs store heat from compression giving a predicted efficiency of 70%—comparable to pumped hydro	• Low roundtrip efficiency (54%) with waste heat from combustion used to heat expanding air—42% without • Very low storage energy density (2.4 kWh/m^3) • Must be located near suitable • geologic caverns		• Approximately 1/3 of output energy is derived from natural gas feed to combustion turbines resulting in additional greenhouse gas emissions

Hydrogen references: Crotogino and Huebner (2008), Denholm and Kulcinski (2004). Batteries: EPRI-DOE (2003). Pumped hydro: Denholm and Kulcinski (2004), Gleick (1994). CAES: Crotogino and Huebner (2008).

For the fuel cell–based electricity storage systems, the roundtrip efficiency is influenced by both the electrolyzer and fuel cell efficiencies. However, the fuel cell efficiency has a greater impact on the overall LCOE. For the aboveground storage system, the fuel cell efficiency has approximately twice the impact on overall cost as the electrolyzer efficiency: 0.54% change in LCOE per percent change in fuel cell efficiency versus 0.23% change in LCOE per percent change in electrolyzer efficiency. The difference is due to the fuel cell's position as the last step in the process chain. Inefficiencies in the fuel cell operation are projected through all of the process stages, whereas inefficiencies in the electrolyzer only affect the costs for the electrolyzer and sensitivity to input electricity price.

4.2. Excess Hydrogen Production for Vehicles

Production of a small amount of excess hydrogen for the vehicle market reduces the overall cost of energy for the scenario by about 6% (hydrogen cost decreases from $4.98/kg to $4.69/kg), but excess hydrogen produced in this way is still not competitive with hydrogen produced in a distributed, dedicated electrolysis process ($3.44/kg). However, for production of larger volumes of hydrogen, the energy storage process is competitive with a dedicated facility. Excess hydrogen is produced in the energy storage scenario for $3.3 3/kg versus $6. 86/kg (untaxed) for a dedicated electrolysis facility at the same production level.

4.3. Non-Economic Benefits and Drawbacks

Many of the benefits and drawbacks listed for each technology in Table 20 have economic impacts. However, these impacts may be difficult to quantify, especially for generic analyses. They are listed here to indicate other considerations that may become important for specific projects.

Efficiency losses in the storage system have an environmental impact because of the net additional electricity that must be generated. If the electricity used to charge the system is from a renewable source, efficiency losses have the effect of reducing the amount of traditionally generated electricity that can be displaced. If the electricity used to charge the system is from non-renewable sources, additional generation is required to compensate for the storage system efficiency losses. The negative emissions impact of the storage system could be partially offset by allowing more efficient operation of traditional power generation systems. Coal-fired power plants usually have lower efficiency (and higher emissions per kWh generated) at lower operating levels. Charging storage systems during periods of low demand could allow these generators to operate at higher production levels and thus reduce their per-kWh emissions.

4.4. Future Analysis Work

The use of hydrogen for energy storage provides unique opportunities for integration between the transportation and power sectors. In regions of high electricity transmission congestion or remote locations currently without transmission lines, transport of hydrogen may prove more economical than expansion of the electric grid. High penetration of renewable electricity generation will also present challenges for grid management. Electricity storage has been investigated as a strategy for integrating large amounts of renewable electricity onto the grid (Denholm and Margolis 2006). Combining energy storage and hydrogen production could provide additional economic and environmental advantages that were not explored in this chapter.

REFERENCES

Afgan, N. H. & Carvalho, M. G. (2004). "Sustainability Assessment of Hydrogen Energy Systems. " *International Journal of Hydrogen Energy, 29(13)*, 1327–1342.

Argonne National Laboratory (2009). Hydrogen Delivery Components model v1. 1. www. hydrogen delivery. html, accessed 2009.

Argonne National Laboratory (2009a). Hydrogen Delivery Scenario Analysis model v2. 02. www. hydrogen delivery. html, accessed.

Borns, D. J. & Lord, A. S. (2008). "Geologic Storage of Hydrogen. " *DOE Hydrogen Program FY 2008 Annual Progress Report*. Washington, DC: U. S. *Department of Energy*.

Casey, E. (2009). "Salt Dome Hydrogen Storage. " Fuel Pathways Integration Technical Team (FPITT) Meeting, Washington, DC, March 7.

Crotogino, F. & Huebner, S. (2008). "*Energy Storage in Salt Caverns:*

Developments and Concrete Projects for Adiabatic Compressed Air and for Hydrogen Storage. " Solution Mining Research Institute Spring 2008 Technical Conference, Porto, Portugal, April 28– 29.

Denholm, P. & Margolis, R. (2006) "Very Large-Scale Deployment of Grid-Connected Solar Photovoltaics in the United States: Challenges and Opportunities. " Preprint to be presented at Solar 2006 Denver, Colorado, July 8–13.

Denholm, P. & Kulcinski, G. L. (2004). "Life Cycle Energy Requirements and Greenhouse Gas Emissions from Large Scale Energy Storage Systems. " *Energy Conversion and Management*, *45*, 2153–2172.

Dennis, R. A. (2008a). "Hydrogen Combustion in Near Zero Emissions IGCC: DOE Advanced Turbine Program. " ASME/IGTI Turbo Expo 2008, Berlin, Germany, June 9–13.

Dennis, R. A. (2008b). "Hydrogen Turbines for Coal-Based IGCC: DOE Advanced Turbine Program. " ASME/IGTI Turbo Expo 2008, Berlin, Germany, June 9–13.

Dhathathreyan, K. S. & Rajalakshmi, N. (2007). "Polymer Electrolyte Membrane Fuel Cell. " S. Basu, ed. *Recent Trends in Fuel Cell Science and Technology*. New York: Springer, 40–1 15.

DOE (2007). Multi-year Research, Development and Demonstration Plan (MYPP). Washington, DC: U. S. Department of Energy

DOE (2009). *Fuel Cells Technologies Program Web site: Types of Fuel Cells, Polymer Electrolyte Membrane Fuel Cells.*

www. eere. energy. gov/hydrogenandfuelcells/fuelcells/fc_types. html#pem, accessed September 2009.

DOE-FE (2005). "10 New Projects to Help Enable Future Turbines for Hydrogen Fuels. " *Fossil Energy Techline*, September 8.

www. fossil. energy. gov/news/techlines/2005/tl_ enabling_turbines_ awards. html. Washington, DC: U. S. Department of Energy Fossil Energy Office.

Electricity Storage Association (2009). Pumped Hydro Web page. www. electricitystorage. org/site/technologies/pumped_hydro, accessed.

EPRI (2003). *Comparison of Storage Technologies for Distributed Resource Applications*. Palo Alto, CA: Electric Power Research Institute.

EPRI (2006). *Technology Review and Assessment of Distributed Energy Resources: Distributed Energy Storage*. Palo Alto, CA: Electric Power Research Institute.

EPRI (2007). *Vanadium Redox Flow Batteries: An In-Depth Analysis*. Palo Alto, CA: Electric Power Research Institute.

EPRI-DOE (2003). *EPRI-DOE Handbook of Energy Storage for Transmission and Distribution Applications*. Palo Alto, CA: Electric Power Research Institute; Washington, DC: U. S. Department of Energy.

EPRI-DOE (2004). *EPRI-DOE Handbook Supplement of Energy Storage for Grid- Connected Wind Generation Applications*. Palo Alto, CA: Electric Power Research Institute; Washington, DC: U. S. Department of Energy.

Gleick, P. H. (1994). "Water and Energy. " *Annual Review of Energy and Environment*, vol. *19*, 267–299.

Greenblatt, J. B., Succar, S., Denkenberger, D. C. & Williams, R. H. (2004). "Toward Optimization of a Wind/Compressed Air Energy Storage (CAES) Power System. " Electric Power Conference, Baltimore, MD, March 30–April 1.

IKA (2009). *Large Hydrogen Underground Storage*. Aachen, Germany: RWTH Aachen University, Institut für Kraftfahrzeuge. www. ika. rwth-aachen. de/index-e. php.

Juste, G. L. (2006). "Hydrogen Injection as Additional Fuel in Gas Turbine Combustor: Evaluation of Effects. " *International Journal of Hydrogen Energy*, *31*, 2112–2121.

Lambert, T., Gilman, P. & Lilienthal, P. (2006) "Micropower System Modeling with HOMER. " In *Integration of Alternative Sources of Energy*, by F. Farret and M. Simões. New York: Wiley. http://homerenergy. com/documentation. asp.

Lipman, T. E., Edwards, J. L. & Kammen, D. M. (2004). "Fuel Cell System Economics: Comparing the Costs of Generating Power with Stationary and Motor Vehicle PEM Fuel Cell Systems. " *Energy Policy*, vol. *32(1)*, 101–125.

LoganEnergy (2007). *NAS Keflavik – Keflavik International Airport PEM Demonstration Project, Final Project Report*. W9 1 32T-05-R-0028. Edinburgh, Scotland: Logan Energy Limited.

LoganEnergy (2008). *U. S. Army Engineer Research and Development Center, Communications Facility, Fort Hood, Texas, Midpoint Project Report*. W9 1 32T-05-C0031. Edinburgh, Scotland: Logan Energy Limited.

Metalprices. com (2009). Vanadium Ferro 80% - Charts. www. metalprices. com/FreeSite/Charts/v_ferro_charts. html?weight=lb# Chart 5, accessed 2009.

Nakhamkin, M. (2007). "*Novel Compressed Air Energy Storage Concepts Developed by ESPC*. " Electrical Energy Storage Systems Applications and Technologies (EESAT) Conference, San Francisco, CA, September 23–26.

Nakhamkin, M., Chiruvolu, M. & Daniel, C. (2007). "*Available Compressed*

Air Energy Storage (CAES) Plant Concepts. " Power-Gen Conference, MN, December.

Oak Ridge National Laboratory (2008). *Bootstrapping a Sustainable North American PEM Fuel Cell Industry: Could a Federal Acquisition Program Make a Difference?* ORNL/TM-2008/183. Oak Ridge, TN: Oak Ridge National Laboratory.

O'Hayre, R., Cha, S., Colella, W. & Prinz, F. (2006). *Fuel Cell Fundamentals.* New York: Wiley.

Phadke, A., Goldman, C., Larson, D., Carr, T., Rath, L. : Balash, P. & Yih-Huei, W. (2008). *Advanced Coal Wind Hybrid: Economic Analysis.* LBNL- 1 248E. Berkeley, CA: Ernest Orlando Lawrence Berkeley National Laboratory.

Pilavachi, P. A., Stilianos, D. S., Vasilios A. P. & Afgan, N. H. (2009). "Multi-Criteria Evaluation of Hydrogen and Natural Gas Fuelled Power Plant Technologies. " *Applied Thermal Engineering*, *29*, 2228–2234.

Ramsden, T., Kroposki, B. & Levene, J. (2008). *Opportunities for Hydrogen-Based Energy Storage for Electric Utilities.* Golden, CO: National Renewable Energy Laboratory.

Schoenung, S. M. & Eyer, J. (2008). Benefit/Cost Framework for Evaluating Modular Energy Storage: A Study for the DOE Energy Storage Systems Program. Livermore, CA: Sandia National Laboratories.

Schoenung, S. M. & Hassenzahl, W. V. (2003). *Long- vs. Short-Term Energy Storage Technologies Analysis: A Life-Cycle Cost Study.* A Study For The DOE Energy Storage Systems Program. Livermore, CA: Sandia National Laboratories.

Siemens Power Generation (2007). *Advanced Hydrogen Turbine Development.* DE-FC26- 05NT42 644. Morgantown, WV: National Energy Technology Laboratory.

Sioshansi, R., Denholm, P., Jenkin, T. & Weiss, J. (2008). "Estimating the Value of Electricity Storage in PJM: Arbitrage and Some Welfare Effects. " Preprint submitted to *Energy Economics*, August, *28*, 2008.

Stevens, J. & Lightner, V. (2003). "Development of a 50-kW Fuel Processor for Stationary Fuel Cell Applications Using Revolutionary Materials for Absorption-Enhanced Natural Gas Reforming (New FY 2004 Project). " *Hydrogen, Fuel Cells, and Infrastructure Technologies FY 2003 Progress Report.* Washington, DC: U. S. Department of Energy.

Stone, H. J. (2005). *Economic Analysis of Stationary PEM Fuel Cell Systems.* Columbus, OH: Battelle Memorial Institute.

van der Linden, S. (2006). "Bulk Energy Storage Potential in the USA,

Current Developments and Future Prospects. " *Energy*, vol. *31(15)*, 3446–3457.

End Notes

[1] The levelized cost is the total annualized cost of the initial capital investment, interest, replacement costs, disposal and/or salvage value, and variable and fixed operating costs over the lifespan of the facility divided by the total yearly energy output from the system.

[2] The levelized cost of energy includes electricity fed to the grid plus hydrogen for vehicles but not hydrogen used as an intermediate energy storage medium. See *http://homerenergy.com/documents/MicropowerSystemModelingWithHOMER.pdf* for a detailed explanation of the cost calculations.

[3] PJM Interconnection is a regional transmission organization that coordinates the movement of wholesale electricity in all or parts of 13 states and the District of Columbia (see www.pjm.com).

[4] Current Forecourt Hydrogen Production from Grid Electrolysis (1,500 kg per day) version 2.1.2 (www.hydrogen

[5] HOMER: https://analysis.nrel.gov/homer.

[6] GDP Implicit Price Deflator (Index, 2000 = 100), available from Short Term Energy Outlook, Table 9a, August 12, 2008, www.eia.doe.gov/emeu/steo/pub/contents.html.

[7] Spectra Energy, Saltville Gas Storage: www.spectraenergy.com/what we do/businesses/us/assets/saltville.

[8] Current Central Hydrogen Production from Grid Electrolysis version 2.1.1 and Future Central Hydrogen Production from Grid Electrolysis version 2.1.1 (www.hydrogen

[9] Current Forecourt Hydrogen Production from Grid Electrolysis (1,500 kg per day) version 2.1.2, 100% debt financing, 0% IRR, 0% inflation, 0% taxes, www.hydrogen

[10] Current Central Hydrogen Production from Grid Electrolysis version 2.1.1, 100% debt financing, 0% IRR, 0% inflation, 0% taxes, www.hydrogen

In: Energy Storage: Issues and Applications ISBN: 978-1-61209-517-2
Editors: Jonathan M. Bowen

Chapter 3

THE VALUE OF CONCENTRATING SOLAR POWER AND THERMAL ENERGY STORAGE*

Ramteen Sioshansi and Paul Denholm

ABSTRACT

This paper examines the value of concentrating solar power (CSP) and thermal energy storage (TES) in four regions in the southwestern United States. Our analysis shows that TES can increase the value of CSP by allowing more thermal energy from a CSP plant's solar field to be used, by allowing a CSP plant to accommodate a larger solar field, and by allowing CSP generation to be shifted to hours with higher energy prices. We analyze the sensitivity of CSP value to a number of factors, including the optimization period, price and solar forecasting, ancillary service sales, capacity value and dry cooling of the CSP plant. We also discuss the value of CSP plants and TES net of capital costs.

1. INTRODUCTION

Recent and ongoing improvements in thermal solar generation technologies coupled with the need for more renewable sources of energy have

---

* This is an edited, reformatted and augmented edition of a National Renewable Energy Laboratory publication, Report NREL-TP-6A2-45833, dated February 2010.

increased interest in concentrating solar thermal power (CSP). Unlike solar photovoltaic (PV) generation, CSP uses the thermal energy of sunlight to generate electricity. Two common designs of CSP plants—parabolic troughs and power towers—concentrate sunlight onto a heat-transfer fluid (HTF), which is used to drive a steam turbine. An advantage of CSP over non-dispatchable renewables is that it can be built with thermal energy storage (TES), which can be used to shift generation to periods without solar resource and to provide backup energy during periods with reduced sunlight caused by cloud cover.[1] The storage medium is typically a molten salt, which has extremely high storage efficiencies in demonstration systems.

Adding TES provides several additional sources of value to a CSP plant. First, unlike a plant that must sell electricity when solar energy is available, a CSP plant with TES can shift electricity production to periods of highest prices. Second, TES may provide firm capacity to the power system, replacing conventional power plants as opposed to just supplementing their output. Finally, the dispatchability of a CSP plant with TES can provide high-value ancillary services such as spinning reserves.[2]

Building a CSP plant with TES introduces several sizing options as the plant essentially consists of three independent but interrelated components that can be sized differently: the power block, the solar field, and the thermal storage tank. The size of the power block is the rated power capacity of the steam turbine, and it is typically measured by the rated input of the power block in megawatts of thermal energy (MW-t) or in the rated output of the power block in megawatts of electric energy (MW-e).

The size of the solar field, in conjunction with solar irradiance, determines the amount of thermal energy that will be available to the power block. The sizing of the solar field is important because the relative size of the solar field and power block will determine the capacity factor of the CSP plant and the extent to which thermal energy will be wasted. Undersizing the solar field will result in an underused power block and a low capacity factor for the CSP plant, because of the lack of thermal energy during all hours but those with the highest solar resource.[3] An oversized solar field, on the other hand, will tend to result in thermal energy being wasted because the power block will not have sufficient capacity to use the thermal energy from the solar field in many hours. The size of the solar field can either be measured in the actual area of the field or by using the solar multiple, which normalizes the size of the solar field in terms of the power block size. A solar field with a solar multiple of 1.0 is sized to provide sufficient energy to operate the power block at its rated capacity under reference conditions. The area of a solar field with a higher or lower solar multiple will be scaled linearly; i.e. a field with a solar multiple of 2.0 will cover twice the area of a field with a solar multiple of 1.0.

The size of storage determines both the thermal power capacity of the heat exchangers between the storage tank and the HTF (measured in MW-t) and the total energy capacity of the storage tank. While the energy capacity of the storage tank can be measured in MWh-t, measuring it in terms of the number of hours of storage is often more convenient. We define the storage capacity as the number of hours that the storage tank can be charged at maximum capacity, which is very similar to the number of hours of discharge capacity because the roundtrip efficiency of the TES is about 98.5%. It is worth noting that another benefit of building a CSP plant with TES is that storage can allow the plant to be built with a larger solar field because excess thermal energy can be placed into storage for use later.

For a merchant CSP developer, the decision to build and the choice of the size of a CSP plant will be governed not only by the amount of solar energy available but also by the pattern (coincidence) of solar resource and by electricity prices. Clearly, high electricity prices and an abundance of solar resource are necessary for CSP to be economic, but a lack of correlation between solar availability and electricity prices can make CSP economically unattractive. TES can improve the economics by shifting generation to higher-priced hours, but this adds capital costs and some efficiency losses in the storage cycle.

In this paper, we examine the value of adding TES to CSP plants in a number of power systems in the southwestern United States. Using a model that optimally dispatches CSP with TES into existing electricity markets, we examine the potential operating profits (i.e., net revenues from energy sales, not accounting for fixed capital costs) that a CSP plant can earn. We show how these profits vary as a function of plant size. We also show the sensitivity of operating profits to different assumptions, including the possibility of selling ancillary services, the optimization process used, and the use dry rather than wet cooling. We show that while the current cost of CSP technologies make them uneconomic based on energy value alone, addition of TES improves the economics of CSP. We also show that when the value of ancillary services and capacity are taken into account, TES and CSP can be economic even with current technology costs.

The remainder of this paper is organized as follows: Section 2 describes our model and the assumptions underlying our analysis. Section 3 presents our analysis of CSP operating profits under the base scenario while Section 4 discusses the sensitivity of those profits to these assumptions. Section 5 discusses the net profitability of TES and CSP plants and Section 6 concludes.

2. CSP Model and Simulations

We model the capabilities and costs of CSP plants using a mixed-integer program (MIP) that is based on the Solar Advisor Model (SAM) developed by the National Renewable Energy Laboratory (NREL)[4] SAM is a software package that can model in detail the hourly energy production of PV, concentrating PV, and CSP plants, with different configurations and in different locations. This production data can be coupled with cost data to compute different economic benchmarks for solar systems, such as the levelized cost of energy. SAM also has some capabilities to optimize the dispatch of a CSP plant with TES by using heuristic "time-ofday" type rules. These heuristics can be used to compare the cost of energy from a CSP plant to other conventional generators.

Unlike SAM, our MIP is formulated to fully optimize the dispatch of a CSP plant to maximize net revenues from energy sales rather than using heuristic rules.[5] The dispatch assumes the CSP plant is a price-taking generator that treats prices as fixed. Because we model only a single CSP plant, this price-taking assumption is reasonable, as the operation of the CSP plant would have at most a marginal impact on the dispatch of other generators. The capabilities and costs of the CSP plants are simulated using the baseline CSP system in SAM, which has a wet-cooled power block with a design capacity of 110 MW-e We simulate a set of CSP plant sizes with solar multiples ranging between 1.5 and 2.7 and between zero and twelve hours of TES.

SAM models the three components of a CSP plant—the solar field, TES system, and power block—on an hourly basis, and we use SAM for developing our CSP dispatch model. The solar field model determines the amount of thermal energy (measured in MWh-t) collected by the field that is available to be put into the TES or power block. This available thermal energy is determined by ambient sunlight as well as solar field size and the efficiencies of the components such as concentrators, collectors, and the HTF used in the solar field. We use this hourly available solar data as an input to our model.

We model CSP plants in four different locations, which are summarized in Table 1. The plants were all simulated using energy price and solar data from

2005. The solar data were obtained from NREL's Renewable Planning Model.[6 7] For the Daggett and Texas CSP plants, hourly real- time energy and day-ahead ancillary service price data from their respective wholesale electricity markets were used. In the case of Daggett, price data from the California Independent System Operator (CAISO) SP1 5 zone is used, whereas the Texas plant uses prices from the Electricity Reliability Council of Texas (ERCOT) western zone. For the locations in Arizona and New Mexico, load lambda data from 2005 for the incumbent utilities—Arizona Public Service (APS) and Public Service New Mexico (PNM), respectively—were used. Load lambda data were obtained from Form 714 filings with the Federal Energy Regulatory Commission.

The TES is modeled as a stock and flow system with losses, which accounts for the loss of thermal energy that is kept in storage over time. Furthermore, energy that goes through the storage cycle will experience some additional losses because of inefficiencies in heat transfer from the solar field to the TES and then to the power block. We assume that hourly energy losses in TES will be 0.03 1%.[8] SAM models efficiency losses from using TES by multiplying the gross output of the power block by a term that is a non-linear function of the fraction of thermal energy delivered to the power block that comes from TES. To maintain linearity of our MIP model, we instead assume that 1.5% of energy taken through the storage cycle will be lost—which roughly approximates the non-linear term in SAM. We assume the TES system is sized to allow the power block to operate at its maximum load using energy from storage alone. For the 110 MW-e power block, this corresponds to a power capacity of approximately 340 MW-t. We also assume that the TES system has the same power capacity for charging and discharging.

Table 1. Location of CSP Plants Studied

CSP Site	Location	Price Data
Arizona	Gila Bend, Arizona (32°57'N, 1 12°57' W)	APS Load Lambda
Daggett	Daggett, California (34°51'N, 116°51' W)	CAISO SP15 Zone
New Mexico	Southern New Mexico (31°39'N, 108°39' W)	PNM Load Lambda
Texas	Western Texas (32°21 'N, 1 02°2 1' W)	ERCOT West Zone

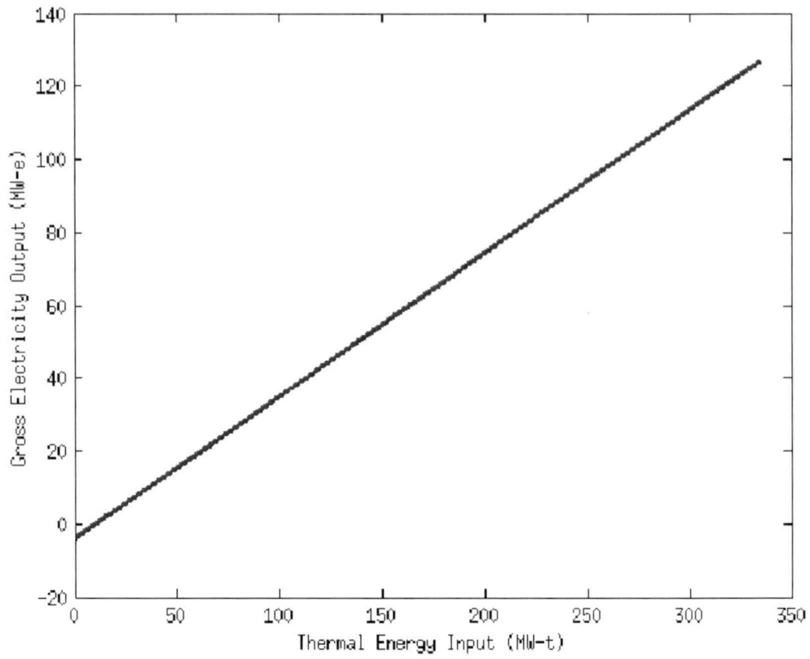

Figure 1. Polynomial heat-rate curve in SAM

The power block is modeled using a heat-rate curve, which gives gross electric output as a function of thermal energy put into the power block. SAM uses a third-order polynomial heat rate curve, which is shown in Figure 1. The figure shows that the polynomial has very little curvature, and for this reason we approximate the curve as a linear function in our dispatch model. As is suggested in the figure, SAM allows the power block to be operated at up to 115% of its design capacity (126.5 MW-e), and we assume that the power block can generate up to this level. We further assume that the power block must be run at a minimum load of 40% of its design point (i.e., gross generation of at least 44 MW-e) whenever the power block is operating. We further require that the power block be online for a minimum of two consecutive hours whenever it is started up, and we assume that 58.3 MWh-t of thermal energy is required to start up the power block, [9] which takes one hour.

The output given by the function in Figure 1 is gross electricity production, and does not account for the parasitics of various components in the CSP plant. These parasitics include energy expended for operating the HTF pumps in the TES system, the cooling tower, and balance of plant. [10] As with the heat-rate curve, these parasitics are represented as third-order polynomials in SAM. Our model combines cooling tower and balance of plant parasitics into a single power block parasitic. The HTF parasitics are modeled separately because they are a function of the amount of energy taken out of TES, while the other parasitics are a function of gross power block generation. Figures 2 and 3 show the polynomial representation of the parasitics in SAM, as well as the piecewise-linear approximations used in our model. SAM also includes an operating cost estimate of $0.70 per MWh-e generated, which is included in the objective function of the dispatch model.

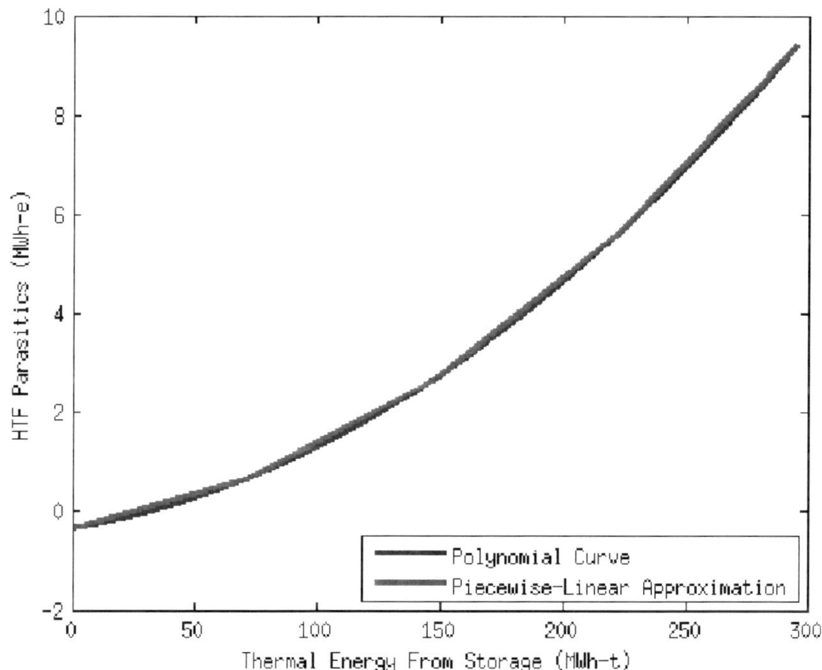

Figure 2. Polynomial representation of HTF parasitics in SAM and piecewise-linear approximation in dispatch model

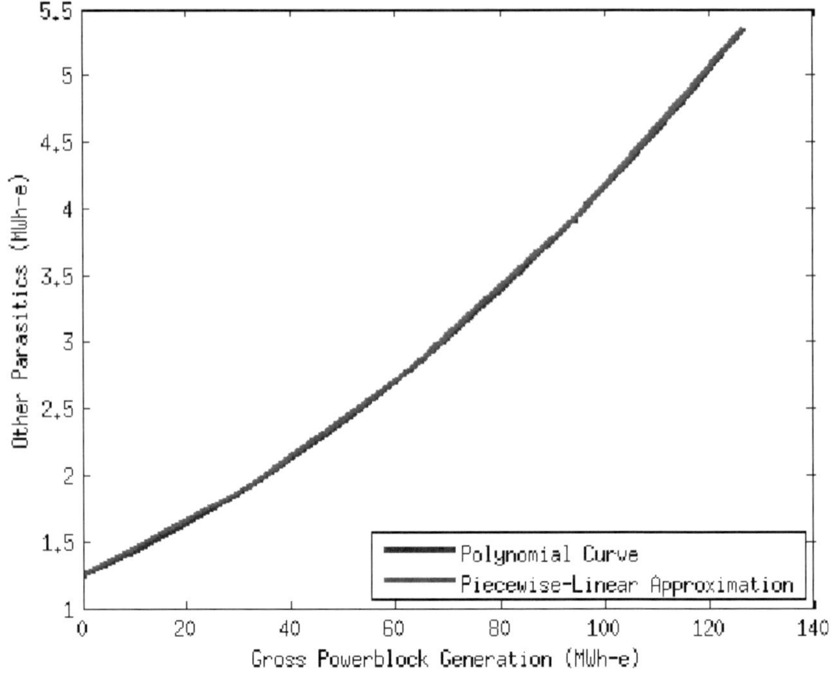

Figure 3. Polynomial representation of power block parasitics in SAM and piecewise-linear approximation in dispatch model

3. Operating Profits of CSP Plants with TES in Energy-Only Markets

As discussed in Section 1, the addition of TES to a CSP plant provides three separate sources of value: energy, capacity, and ancillary services. Our analysis first considers the case of a CSP plant used solely to sell energy into the wholesale market. As discussed in Section 2, we use real- time energy price data from the CAISO and ERCOT market for the Daggett and Texas CSP plants, and we use load lambda data for CSP plants in Arizona and New Mexico, where restructured wholesale markets do not exist. Our analysis first considers the case of a CSP plant operated with perfect foresight of future solar availability and energy prices. Following Sioshansi et al. (2009) the optimization is done one day at a time using a rolling two-day planning horizon. Thus, we assume that dispatch decisions are made one day at a time; however, the next day's energy prices and solar availability are accounted for in making dispatch decisions each day. This use of a two-day planning horizon ensures that energy in TES at the end of the day has carryover value. Without this use of a two-day optimization horizon, energy would never be kept in TES at the end of a day because it would have no value. Figure 4 shows an example dispatch of a CSP plant at the Texas site with six hours of TES and a solar multiple of 2.0 over the course of a winter day, along with available energy from the solar field and hourly energy prices. The figure shows a typical winter price profile, which has highest demand and prices at the beginning and end of the day. The figure shows that the dispatch follows expected patterns, using energy storage for both the morning and evening demand and price peaks. For instance, in the morning (hours eight through 10) the power block is operated with energy from TES, which was carried over from the previous day to catch the high early-morning prices. The power block is shutdown in hours 15 and 16, when energy prices are relatively low, and the solar field energy is put into TES. The plant then provides energy in the evening peak, using both energy from the solar field and energy from storage.

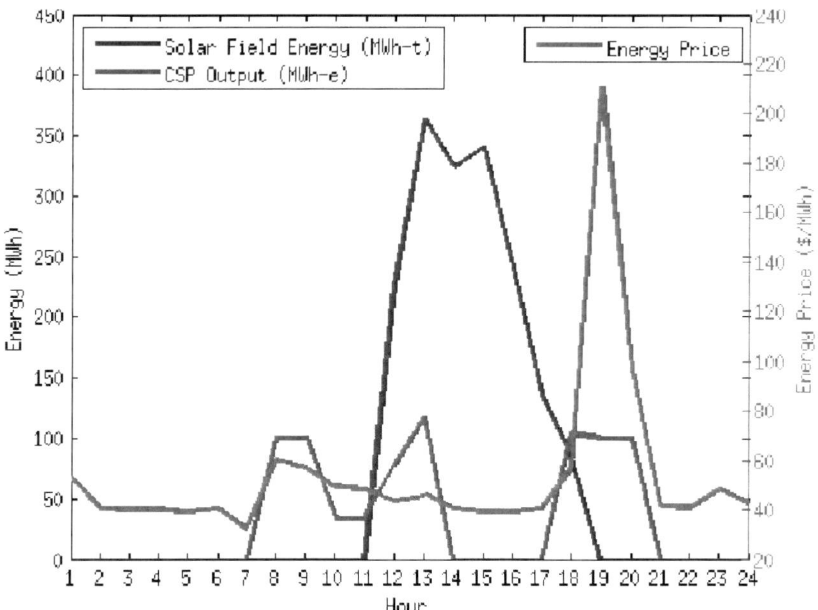

Figure 4. Sample dispatch of a CSP plant with six hours of TES and a solar multiple of 2.0 at Texas site

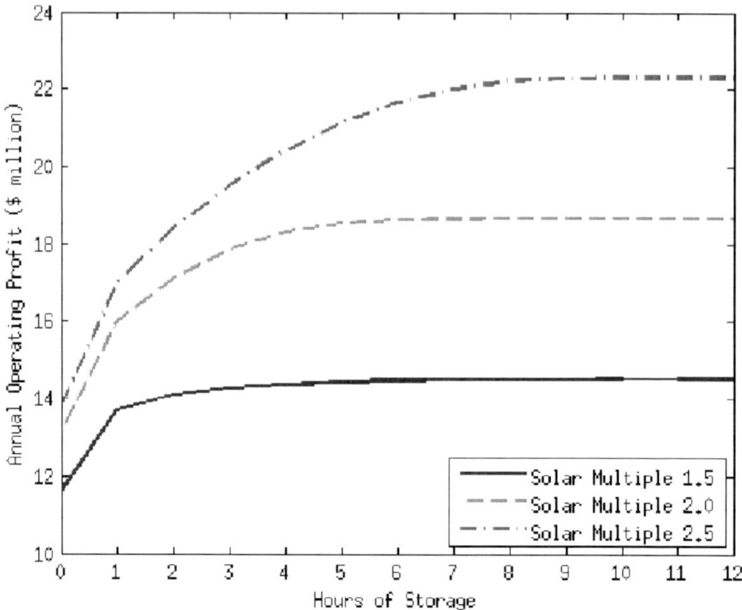

Figure 5. Annual operating profits of a CSP plant at Arizona site

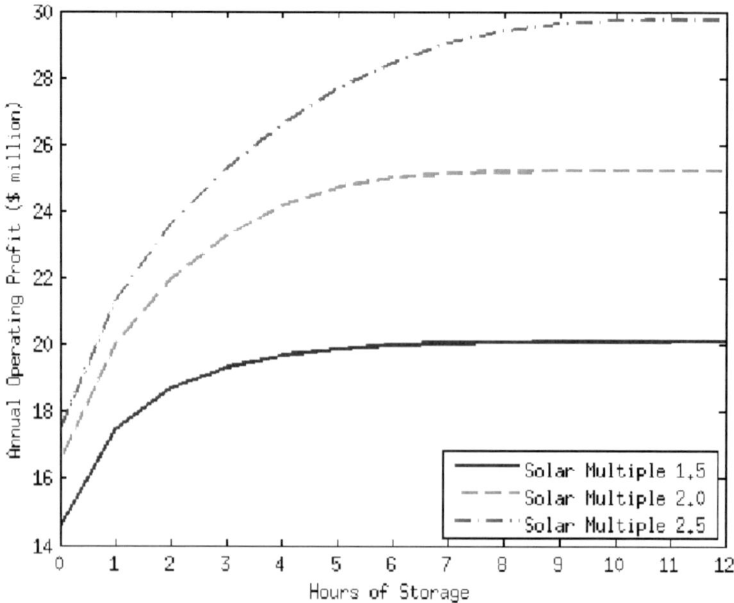

Figure 6. Annual operating profits of a CSP plant at Daggett site

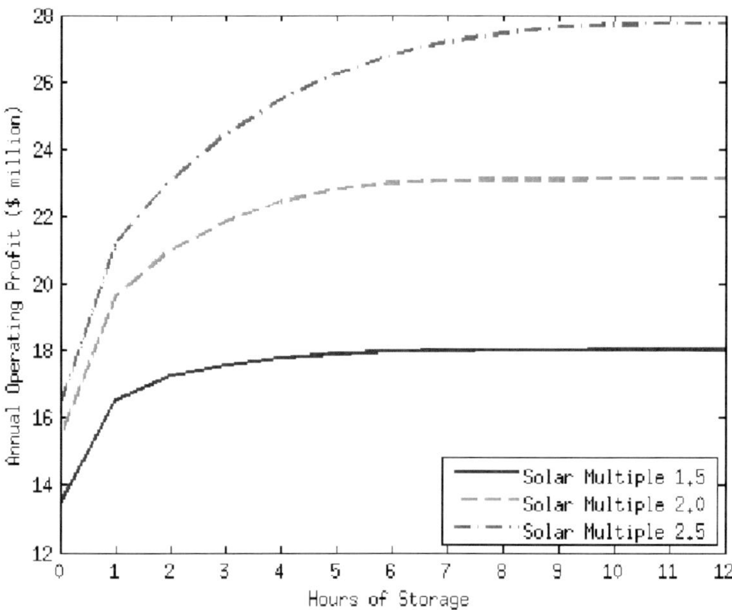

Figure 7. Annual operating profits of a CSP plant at New Mexico site

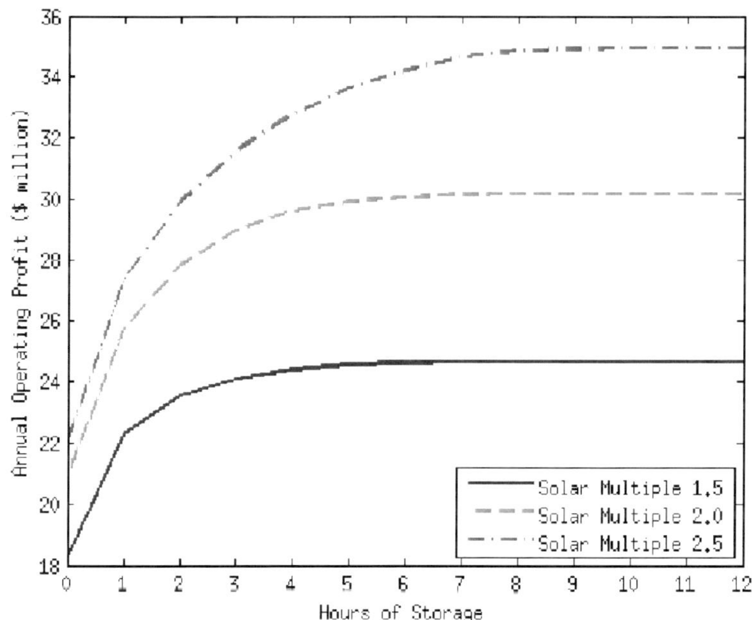

Figure 8. Annual operating profits of a CSP plant at Texas site

Table 2. Solar Energy Available and Average Price of Energy for Different CSP Sites

			Average CSP Selling Price ($/MWh)		Annual Operating Profits ($ million)	
CSP Site	**Solar Field Energy with Solar Multiple 2.0 (GWh-t)**	**Average Energy Price ($/MWh)[11]**	**Solar Multiple 1.5, No TES**	**Solar Multiple 2.0, Six Hours TES**	**Solar Multiple 1.5, No TES**	**Solar Multiple 2.0, Six Hours TES**
Arizona	1,150	41.2	47.0	50.5	11.6	18.6
Daggett	1,181	55.9	58.5	67.9	14.5	25.0
New Mexico	1,088	57.3	61.2	66.2	13.5	23.0
Texas	961	66.4	89.4	98.4	18.2	30.1

Figures 5 through 8 show the operating profits of CSP plants with different-sized solar fields and different amounts of TES at the four CSP sites described in Table 1. The figures highlight the fact that the value of a CSP plant can vary significantly by location, with a CSP plant at the Arizona site earning about 60% of the operating profits of the Texas plant. The figures also show how the operating profits vary with plant size. At all of the sites, the value of TES tapers off at about six to eight hours of storage. Moreover, we see that increasing the size of the solar field only yields noticeable profit increases if sufficient TES is available to shift the solar resource to periods with less available sunlight.

By comparing the average electricity price and the total annual thermal energy produced by different-sized CSP plants, Table 2 shows the underlying cause of the differences in the operating profits of the CSP plants at the different sites. The table shows that the price of energy tends to be more important in determining the profitability of a CSP plant than the amount of solar energy available. Indeed, the Texas site produces the least amount of solar energy, yet the relatively high price of electricity makes it the most profitable site. The table also shows that because of the coincidence of solar insolation with normal load patterns, CSP without TES is between 7% and 35% more valuable than the average price of electricity in the cases evaluated. The table further shows that adding TES increases this value by an additional 7% to 16%.

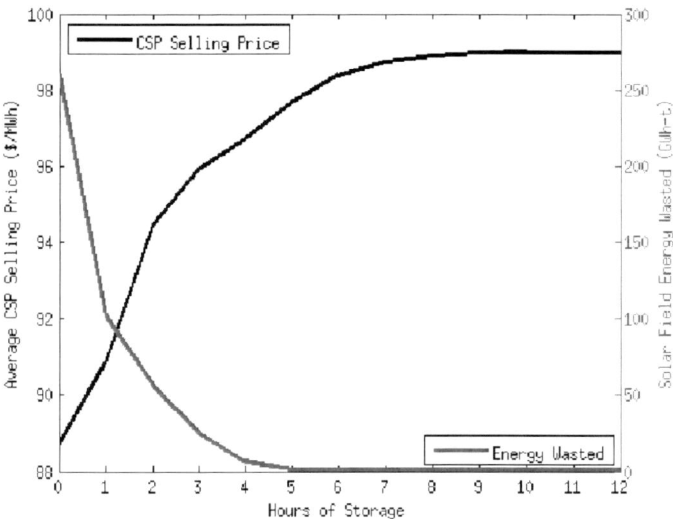

Figure 9. Average selling price of energy ($/MWh) and solar field energy wasted (GWh-t) for a CSP plant at Texas site with solar multiple of 2.0.

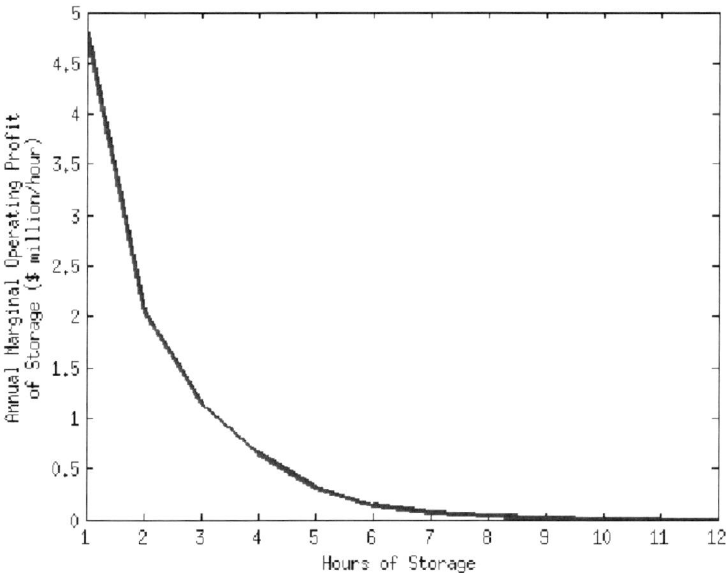

Figure 10. Marginal annual value of each incremental hour of storage for a CSP plant at Texas site with solar multiple of 2.0.

The energy-related value of TES is actually derived from two sources. First, TES allows more of the energy collected by the solar field to be used by placing excess energy that would overload the power block into TES for future use, which then allows a larger solar field to be used with a power block. The second value of TES is that it allows generation to be shifted from periods with lower prices to those with higher prices. Figure 9 demonstrates these two effects for a CSP plant in Texas with a solar multiple of 2.0. It shows that as TES is added to the CSP plant, more solar field energy is able to used, with this benefit of TES flattening at about four hours of storage. Moreover, the figure shows that adding TES helps increase the average selling price of energy from the CSP plant, because the TES allows generation to be shifted between hours. This use of TES increases the average selling price of energy from the CSP plant by about $5/MWh with three hours of storage. Figure 10 summarizes these two effects of adding TES to a CSP plant by showing the marginal annual value of each incremental hour of storage. As suggested by Figure 9, the first two hours of TES are highly valuable because of both the generation-shifting and the increased use of solar energy, with the marginal value of storage nearly flattening after eight hours of TES; this is because with this amount of storage the generation-shifting capability of TES has been exhausted and nearly all of the energy from the solar field can be used.

4. Sensitivity of CSP Profits to Base Case Assumptions

The results presented thus far represent a base case with a set of assumptions regarding the optimization horizon, perfect foresight of solar availability and energy prices, the markets in which the CSP plant participates, and the availability of water for wet cooling of the power block. We now examine the effect of relaxing these base assumptions on the profitability of CSP and the value of TES.

4.1. Impact of Optimization Horizon

The results presented in Section 3 assume that the CSP plant would be dispatched using a rolling one-day optimization period (we also assume that each one-day planning problem would account for the second day). These assumptions allow generation to be shifted between hours within a day, and also allow energy to be stored at the end of each day in anticipation of energy prices the following day. However, these assumptions do not account for the fact that a CSP plant may store energy in anticipation of prices multiple days in the future. For example, Sioshansi et al. (2009) describe a "weekend effect," in which more energy tends to be stored over weekends in anticipation of higher energy prices on weekdays. Considering this weekend effect, our assumption of a one-day planning horizon may understate the potential profitability of a CSP plant, because it does not fully allow for inter-day generation shifting.

We examine the sensitivity of CSP profits by comparing our base case with a one-day planning horizon to one with a rolling one-week planning horizon. As with the one-day model, the weeklong model assumes that the CSP plant has perfect foresight of solar availability and energy prices, and we use an eight-day horizon in the dispatch model to ensure that energy in TES has carryover value at the end of each week. Figure 11 shows the increase in annual operating profits if a CSP plant at the Texas site uses a weeklong planning horizon as opposed to daylong one in its dispatch optimization. The profit increases are given as a percentage of the profits from using a daylong planning period. The figure shows that profits are largely insensitive to the optimization period, with less than a 2.4% increase in profits from switching to a weeklong optimization period, implying that most of the generation shifting with TES is done either within or between adjacent days. The other sites show even less sensitivity to the optimization horizon—with a maximum of between a 1.2% increase in profits from using a weeklong optimization horizon at the Arizona site and a 1.3% increase in profits at the New Mexico site. Also interesting is that the profit increase when using a weeklong optimization period is greatest for a CSP plant configuration with a low solar multiple and more hours of storage. The reason for this result is that with a lower solar multiple, the TES is used less for storing excess solar field energy because there are fewer hours in which the capacity of the power block is binding. Thus, a CSP plant with such a configuration uses TES primarily for shifting generation to higher- priced hours, which will be more sensitive to the planning horizon used.

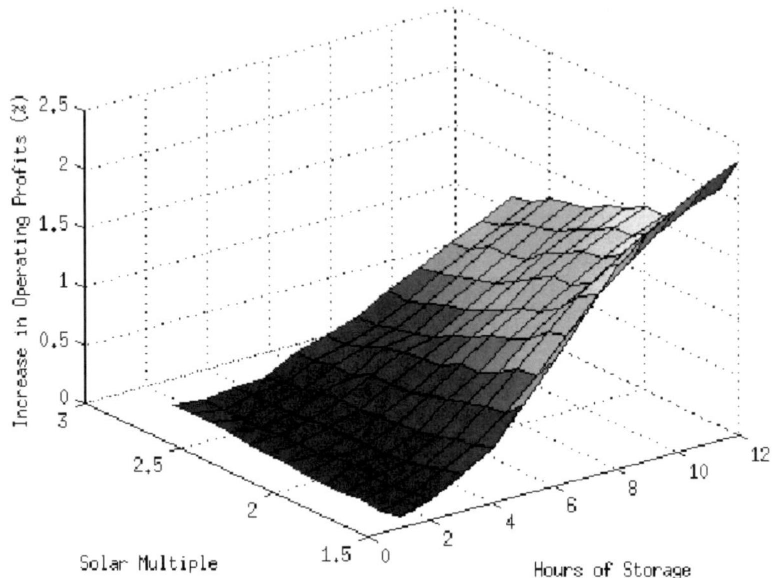

Profit increases are given as a percentage of profits with daylong optimization.

Figure 11. Increase in annual operating profits of a CSP plant at Texas site from using weeklong as opposed to daylong optimization period.

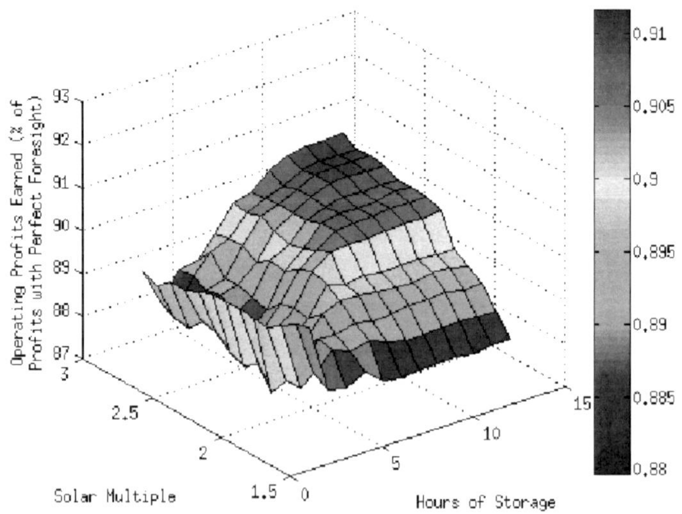

Profits given as a percentage of profits attainable with perfect foresight.

Figure 12. Annual operating profits earned by a CSP plant at Texas site using daily backcasting technique.

4.2. Impact of Solar Availability and Energy Price Forecasting

A major assumption of our base case is that the CSP plant will have perfect foresight of future energy prices and solar availability in conducting its optimization, whereas in practice, storage operators will have a forecast of this data to use. Clearly, the profitability of a CSP plant will be closely related to the quality of forecasts that are available to CSP operators, which may vary based on the type of forecasting models used. Rather than trying to approximate the effectiveness of different forecasting techniques in optimizing the dispatch of a CSP plant, we use the "backcasting" technique described in Sioshansi et al. (2009). This backcasting technique assumes that the CSP plant will be dispatched using historical data only. More specifically, we assume the operation of the CSP plant is optimized one hour at a time using a rolling one-day planning horizon. In hour *t*, the dispatch of the CSP plant is optimized using solar availability and price data from the previous 24 hours, assuming that those price and solar patterns will exactly repeat themselves. This daylong dispatch is used to determine the operation of the plant in hour *t*, and the process is repeated. Once the dispatch of the CSP plant is determined, actual price and solar availability data are used to determine the plant's revenues.

Figure 12 compares the operating profits of a CSP plant at the Texas site using the backcasting technique to those achievable with perfect foresight. The results demonstrate that for all CSP plant sizes considered, this backcasting technique earns at least 87% of the profits that are theoretically attainable with perfect foresight. We conducted the same comparison with the Daggett site, which performed better by earning at least 89% of the perfect-foresight profits.

The relatively good performance of the backcasting technique is not entirely surprising, given that energy prices and solar availability have rather predictable diurnal patterns (Sioshansi et al. 2009). Energy prices tend to peak midday or in the afternoon, and solar availability follows a similar pattern with a midday or afternoon peak (although the peak energy price tends to lag the solar peak by several hours depending on the season). Moreover, while energy prices and solar availability can differ on longer-term bases—for instance, because of seasonal or even annual differences—the use of data from the previous twenty-four hours will tend to capture these effects. Finally, it is important to note that the backcasting technique we examine here does not incorporate any solar or price forecasting in operational decision making. Because weather and prices tend to be fairly predictable, especially on a short-term basis, the value-capture numbers in Figure 12 should be viewed as a lower bound on what can be achieved with the use of forecasting techniques.

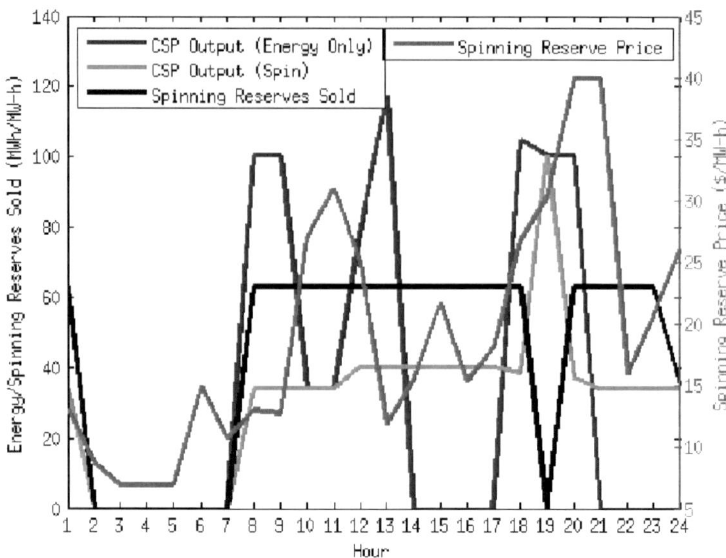

Figure 13. Sample dispatch of a CSP plant with 8 hours of TES and a solar multiple of 2.4 at Texas site when energy and spinning reserve sales are co-optimized.

4.3. Impact of Ancillary Service Sales

The analysis thus far has assumed that the CSP plants will provide only energy, whereas in practice they could also provide ancillary services. Ancillary services typically include regulation, spinning, and non-spinning reserves. Regulation is capacity from a generator that is online, generating, and can quickly be ramped up or down to counterbalance supply deviations on very short, second-long time scales. Regulation is typically deployed using automated controls and is used to maintain supply balance between real-time market settlement periods. Spinning reserve is capacity from a generator that is synchronized with the system, and can provide energy within a longer time frame than regulation does—typically up to thirty minutes. A non-spinning reserve is capacity that can be brought online within a longer time frame than a spinning reserve—typically within half to a full hour—and is deployed to deal with prolonged generator or transmission outages.

Although these capacity services are used in all of the regions that we consider in our analysis, spinning reserve price data are only available from the CAISO and ERCOT markets. Thus, we only consider spinning reserve sales for the Daggett and Texas CSP sites. We assume that the CSP plant will not provide regulation services or non-spinning reserves because of limitations in the rate at which the power block and TES can be started and ramped. We consider two cases in which the plant can provide either 25% or 50% of its generating capacity in the form of spinning reserves. In both cases, we limit the CSP plant such that the total of its energy and spinning reserves sales must be within the power and energy capacity of the plant. These restrictions ensure that if the plant's spinning reserves are deployed in real-time, it can feasibly serve that load in addition to its energy sales. Because spinning reserves are deployed very infrequently (Kempton and Tomić 2005), we do not explicitly model the probability of spinning reserves being deployed or the revenues from that energy being sold.

Figure 13 shows the dispatch of the CSP plant at the Texas site with a solar multiple of 2.0, six hours of TES, and the ability to sell up to 50% of its capacity in spinning reserves over the same one-day period shown in Figure 4 and contrasts it with CSP operations without spinning reserve sales. Several differences in CSP operations can be noted. For example, the power block is run in hours 15 and 16 in order to allow spinning reserves to be sold. In hours eight through 13 and in hours 18 and 20, the CSP plant produces less energy than the case without spinning reserves sales in order to allow it to sell some spinning reserves.

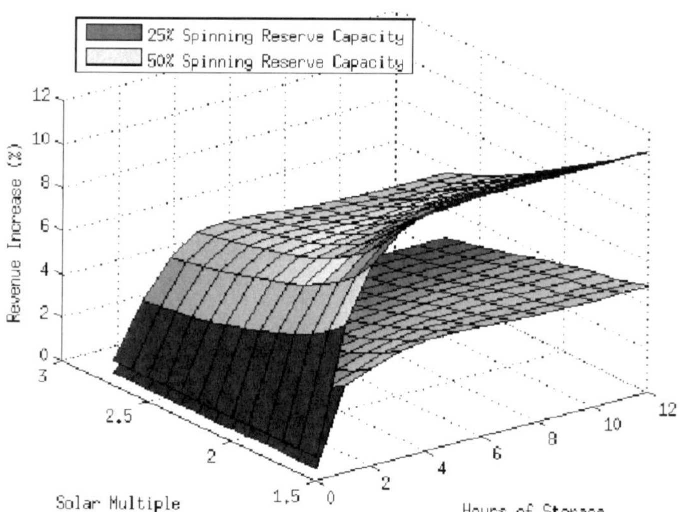

Increase in profits is given as a percentage of profits with energy sales alone.

Figure 14. Increase in annual operating profits of a CSP plant at the Daggett site if spinning reserves can be sold.

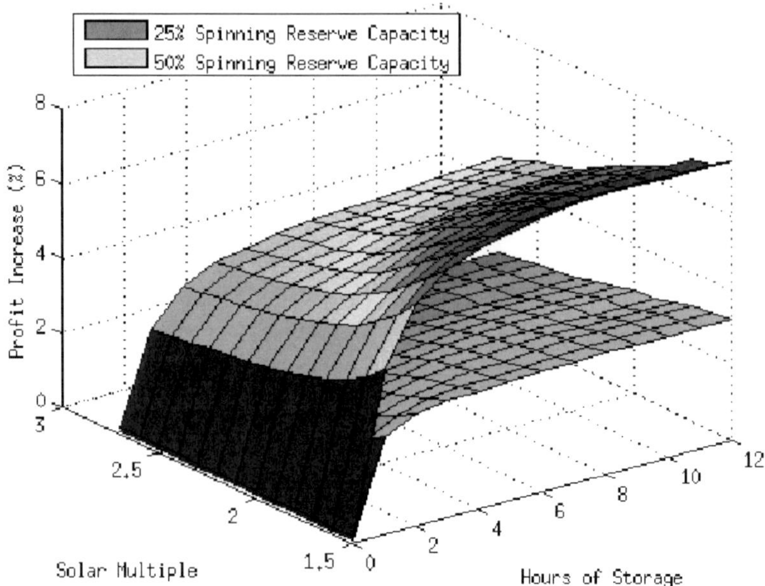

Increase in profits is given as a percentage of profits with energy sales alone.

Figure 15. Increase in annual operating profits of a CSP plant at the Texas site if spinning reserves can be sold.

Figures 14 and 15 summarize the effect of allowing the CSP plants to sell spinning reserves on the operating profits of the Daggett and Texas sites, respectively. The figures show the increase in annual profits from the sale of spinning reserves, as a percentage of the profits with energy sales alone. The figures show that, depending on the plant size, these profit increases can be nontrivial. As can be expected, spinning reserve sales are more valuable when the CSP plant is less constrained in how much it can provide.

4.4. Impact of Capacity Credit

CSP plants, especially those with TES, will generally provide capacity to the system, which reduces the need for other capacity to be built or procured by the utility, load-serving entity, or system operator. Properly valuing this capacity can pose some difficulties in electricity markets, however, as capacity is not necessarily priced in the market and market distortions can suppress these prices. In theory, restructured energy-only markets—such as the CAISO or ERCOT markets—value capacity through scarcity pricing, whereby prices increase with loads to reflect the cost of capacity constraints. Thus, in such a market, the value of capacity should be captured in energy prices. In practice, however, price caps and other regulatory interventions in the market tend to suppress energy prices, and as a result, capacity may not be properly priced in the market. 12 Vertically integrated markets, such as those in Arizona and New Mexico, pose an even greater problem in estimating the value of capacity, because the load lambda data we use as proxies for energy prices do not capture any of the value of capacity.

Because capacity is not properly valued in the energy prices used in our analysis, we use the capital cost of a new gas combustion turbine to compute a proxy for the value of capacity. US DOE (2008) estimates the current capital cost of a combustion turbine to be $625/kW, which translates to a cost of $68.75 million for a 110 MW turbine. We can translate this capital cost of a combustion turbine into an annual cost of capacity by using an annual capital charge rate (CCR) that captures all financing parameters and other costs that would go into building such capacity. Following Denholm and Sioshansi (2009), we assume an 11% CCR, which gives an annual cost of $7.56 million for a 110 MW combustion turbine. If the availability of a CSP plant equals that of a combustion turbine, this value of $7.56 million represents the capacity value of a CSP plant, which represents a 25%-41% increase in the annual value of a CSP plant with TES compared to the base values in Table 2.

In practice, the annual capacity value of a CSP plant would be some fraction of this $7.56 million, depending upon the actual availability of the CSP plant. The availability of a CSP plant will depend on many factors, such as its scheduled and forced outage rate, and the availability of thermal energy from the solar field and TES. While solar field energy is fixed, the availability of energy from the TES will depend upon the dispatch of the plant. Depending on whether the CSP plant is being given a capacity payment and how those payments are determined, the dispatch of the plant could be altered to maximize the sum of energy and capacity revenues. Further work is needed to assess the actual capacity value of CSP plants as a function of location and storage size.

4.5. Impact of Power Block Dry Cooling

Our analysis thus far has assumed that the CSP plants will use a wet-cooled power block. Because of the arid climate of the southwestern United States, the net effect of dry cooling on the profitability of the CSP plants should be considered. Dry cooling the power block tends to increase cooling tower energy losses, with the efficiency reduction depending upon the ambient temperature. This decrease in efficiency tends to increase the value of TES because high ambient temperatures (which give greater efficiency losses with dry cooling) will tend to be correlated with the availability of solar energy. Without TES, the solar energy must be put into the power block, which will produce less energy because of cooling losses. With TES, however, energy from the solar field can be stored and generation can be shifted to hours later in the day when ambient temperatures are reduced. SAM accounts for dry cooling energy losses by multiplying gross power block output by a term that is a fourth-order polynomial function of the ambient temperature. We model dry-cooling energy losses in the same manner, and we use the correction factors from SAM as an input to our dispatch model.

Figure 16 shows the annual operating profit loss from using a dry- as opposed to wet-cooled CSP plant at the Texas site, and Figure 17 summarizes the effect at the four sites for CSP plants with a solar multiple of 2.0. As the two figures show, dry cooling can have a noticeable effect on CSP profits, especially for smaller-sized plants. The figures show that the profit losses are highest for CSP plants without TES (or with very little TES). This greater profit reduction is because with TES, generation can be shifted to hours with lower efficiency losses from dry cooling (i.e., hours with lower ambient temperatures). Figure 18 further demonstrates this phenomenon by showing the reduction in net electrical generation of dry-cooled CSP plants with a solar multiple of 2.0 at the four sites as a function of hours of TES. This figure show that adding an hour of storage results in nearly a 0.5% increase in output from a dry-cooled CSP plant, with incremental storage beyond the first hour giving further increases. One concern with this use of TES in a dry-cooled CSP plant is that market operators may view the shifting of generation to periods with lower ambient temperatures as an attempt by the CSP operator to exercise market power and manipulate energy prices by withholding capacity. While some market monitors may have such a view, this use of TES can be likened to the withholding of water by a hydroelectric or other energy-limited generator, as is described by Lu et al. (2004).

5. Break-Even Cost of CSP Plants

Our analysis thus far has only considered operating profits (i.e. profits from energy or ancillary service sales less variable operating costs). A complete analysis of the value of CSP and TES must trade operating profits against the capital costs of the plant. This analysis would in turn, require (1) multiple years of energy price and solar data to determine the operating profits of the plant over its lifetime and (2) knowledge of the present and future costs of CSP plants with various amounts of energy storage.

The price of large-scale CSP is still highly uncertainty because of large fluctuations in commodities prices and the potential for substantial cost reductions from engineering and manufacturing improvements. Furthermore, the overall cost competitiveness of CSP will depend on changes in fuel prices and carbon policies. Instead of evaluating the overall economics of CSP in this analysis, we examine the cost competitiveness of the TES component. This type of analysis can tell us whether the additional benefits of TES found in this analysis exceed current and future cost estimates. Moreover, if the benefits of TES are seen to exceed these cost estimates, the value of TES above its cost can help to increase the relative cost competitiveness of the entire CSP plant.

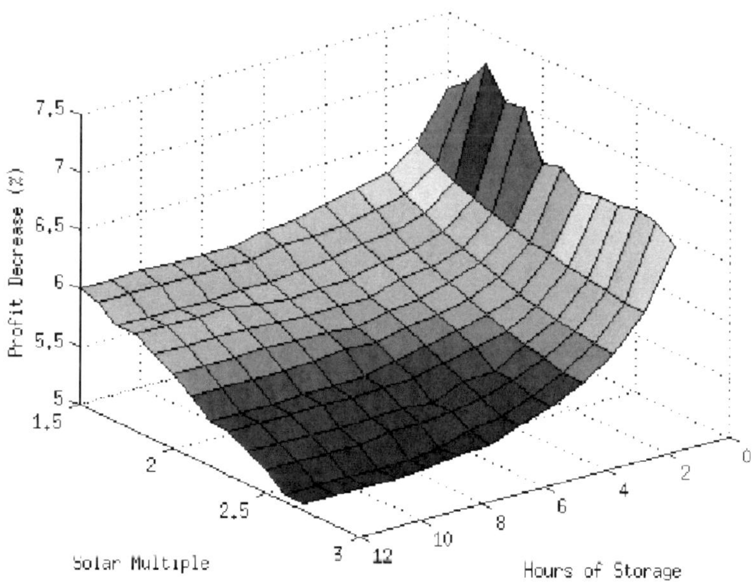

Profit reductions are given as a percentage of profits with wet cooling

Figure 16. Annual operating profit reduction of a dry-cooled CSP plant at Texas site.

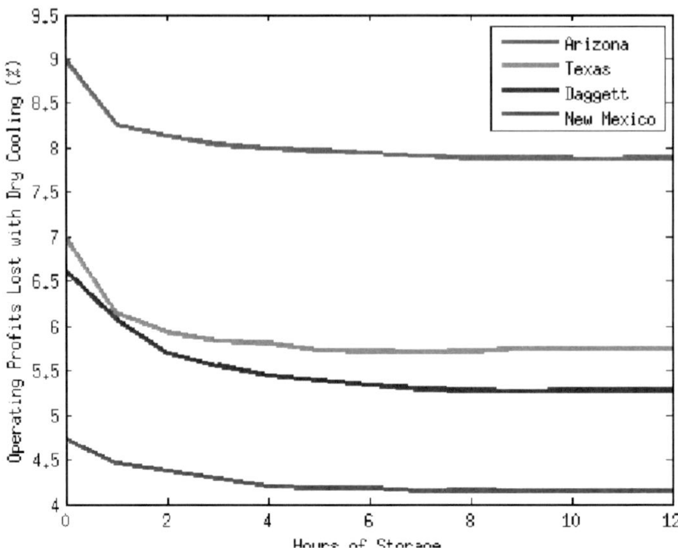

Profit losses given as a percentage of profits with wet cooling.

Figure 17. Net operating profits lost from dry- as opposed to wet-cooled CSP plants with a solar multiple of 2.0.

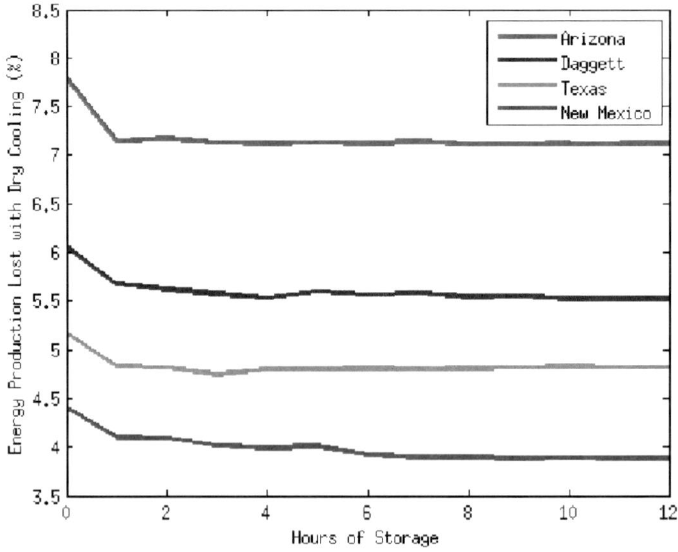

Production losses given as a percentage of net energy production with wet cooling.

Figure 18. Net energy production lost from dry- as opposed to wet-cooled CSP plants with a solar multiple of 2.0.

Stoddard et al. (2006) present estimates of present and future TES costs in 2005 dollars, which can be directly compared to the operating profits estimated thus far, because these estimates used 2005 energy price and load lambda data. We use the 2009 cost estimates for present costs and the 2015 estimates for future costs. We assume that without TES, the solar field has a solar multiple of 1.5 and with TES it has six hours of storage and a solar multiple of 2.0, which reflects the fact that with TES the size of the solar field can be increased relative to the power block. We scale the costs of the TES and the solar field because the analysis in Stoddard et al. (2006) assumes a 100 MW-e power block for the 2009 CSP plant and a 200 MW-e power block for the 2015 CSP plant, whereas we assume a 110 MW-e power block throughout our analysis. These assumptions give an estimated present cost of $156.3 million for TES (plus the additional solar field cost) and a cost of $117.6 million in the future. We also assume that the CSP plant will be eligible for an investment tax credit (ITC) of either 30% or 10%. The 30% reflects the current ITC for solar generators, whereas the 10% reflects a potential future scenario in which the ITC is reduced.

Table 3 summarizes the year-1, break-even cost and return on investment of the TES component of the CSP plant. The break-even cost is defined as the maximum overnight cost of TES, which is justified by the increase in year-1 operating profits of the CSP plant (using an 11% CCR). The return on investment is defined as the percentage of the annual cost of the TES component (based on the cost estimates from Stoddard et al. (2006) and an 11% CCR) that is recovered by the increase in year-1 operating profits of the CSP plant. Thus, a return of investment of 100% or greater indicates that the TES "pays for itself" through higher energy revenues. Clearly, whether the increased revenues from TES will justify the cost of TES will be highly sensitive to the site of the CSP plant, ITC rate, and cost of the TES, but substantial cost reductions appear to be necessary to justify the addition of TES based on energy sales alone.

Table 3 does not include the non-energy value of CSP and TES, especially the value of more firm capacity. As discussed in Section 4, the operating profits of CSP can potentially be increased by ancillary service sales or a capacity credit. Table 4 summarizes the break-even cost when accounting for these sources of revenues. The spinning reserves sales case assumes that 50% of the CSP plant's power capacity can be sold as spinning reserves so long as sufficient energy is available from the solar field or TES. For the Arizona and New Mexico sites, for which we do not have spinning reserve price data, we scale up the break-even cost from the costs in Table 3 based on the increase in the break-even cost of the Daggett and Texas sites—which is a roughly 16% increase. The capacity credit case assumes that the CSP plant has the same availability as a combustion turbine, which gives an annual capacity value of $7.56 million. In practice this availability may be lower, which would reduce the break-even cost, but we use this value to provide an upper bound.

Comparing Table 4 to Table 3 and the cost estimates above shows that TES justifies a greater cost with spinning reserves sales and capacity value. With the 30% ITC, the revenues from TES are greater than its present cost at the Daggett and Texas sites with spinning reserves sales alone. If the capacity credit is included, TES is justified at all sites.

Table 3. Year-1 Break-Even Cost and Return on Investment of TES Using Base Case Profit and Cost Assumptions

CSP Site	Break-even Cost ($ million)		Return on Investment (%)			
			Present Cost		Future Cost	
	30% ITC	10% ITC	30% ITC	10% ITC	30% ITC	10% ITC
Arizona	90.9	70.7	58.1	45.2	77.3	60.1
Daggett	136.4	106.1	87.3	67.8	115.9	90.2
New Mexico	123.4	96.0	78.9	61.4	104.9	81.6
Texas	154.6	120.2	98.9	76.9	131.4	102.2

Table 4. Year-1 Break-Even Cost of TES ($ million) with Spinning Reserves Sales and Capacity Credit

	Spinning Reserves Sales		Spinning Reserve Sales and Capacity Credit	
CSP Site	**30% ITC**	**10% ITC**	**30% ITC**	**10% ITC**
Arizona	105.2	81.8	203.5	158.2
Daggett	160.7	125.0	259.0	201.4
New Mexico	142.8	111.1	241.1	187.5
Texas	176.0	136.9	274.2	213.3

Table 5. Return on Investment of TES with Spinning Reserves Sales and Capacity Credit and a 10% ITC

CSP Site	Return on Investment (%)	
	Present Cost	Future Cost
Arizona	100.9	134.1
Daggett	125.4	166.7
New Mexico	118.7	157.3
Texas	136.4	181.3

Table 5 provides the return on investment of the TES components, considering the ability to provide both spinning reserves sales and capacity credit. In this case, only a 10% ITC is evaluated, and in all cases, the return on investment (ROI) is greater than 100%, meaning that the value of TES is greater than its incremental capital cost. As a result, the incremental value above the cost of TES will improve the cost competitiveness of the entire CSP plant.

6. Conclusions

This paper presented a detailed analysis of the value of CSP plants with TES in for regions in the southwestern United States. Our results showed how operating profits of CSP plants vary as a function of plant size and location. Locational profit differences were shown to be mainly due to differences in energy prices in the different system operator and utility systems considered, with differences in solar resource being a smaller determinant of CSP value. We showed that TES could increase CSP value both by allowing generation to be shifted to higher-priced hours and by increasing the use of thermal energy from the solar field.

We also showed the effect of relaxing several of our base assumptions on CSP profitability. The operating profits of the CSP plant seemed relatively insensitive to the assumption that TES dispatch decisions are made a day at a time, which suggests that most generation shifting is done either within a day or between adjacent days. Longer-term generation shifting over the course of the week and the so-called “weekend effect” were not noticeably apparent for CSP plants, except for plants with small solar fields and large amounts of storage. We also demonstrated that the perfect foresight of energy prices and solar availability assumed throughout our analysis is a relatively good approximation of actual CSP operations; at least 87% of the profits with perfect foresight can be attained by using a very simple backcasting technique. We expect that including price and weather forecasts in the dispatch process can further improve the profitability of a CSP plant. We also showed that adding ancillary services sales can increase CSP profits, and we examined the effect of dry cooling—showing that TES can reduce some efficiency losses associated with dry cooling by shifting generation to hours with lower ambient temperatures.

Despite these benefits of CSP, our analysis suggests that with current capital costs, TES cannot be economically justified on energy value alone. However, including the value of ancillary service sales and capacity can increase the cost competitiveness of TES, making it economic in a number of cases and improving the overall economics of CSP.

References

Denholm, P. & Sioshansi, R. (2009). "The Value of Compressed Air Energy Storage with Wind in Transmission-constrained Electric Power Systems." *Energy Policy 37(8)*, 3149- 3158. doi:10.1016/j.enpol.2009.04.002

Gilman, P., Blair, N., Mehos, M., Christensen, C., Janzou, S. & Cameron, C. (2008). Solar Advisor Model User Guide for Version 2.0. NREL/TP-670-43704. Golden, CO: National Renewable Energy Laboratory.

Kempton, W. & Tomić, J. (2005). "Vehicle-to-grid Power Fundamentals: Calculating Capacity and Net Revenue." *Journal of Power Sources, 144(1)*, 268-279. doi: 10.101 6/j .jpowsour.2004. 12.025

Lu, N., Chow, J. & Desrochers, A. (2004). "Pumped-Storage Hydro-Turbine Bidding Strategies in a Competitive Electricity Market." *IEEE Transactions on Power Systems, 19(2004)*, 834-841. doi:10.1 109/ TPWRS.2004.82591 1

Sioshansi, R., Denholm, P., Jenkin T. & Weiss, J. (2009). "Estimating the Value of Electricity Storage in PJM: Arbitrage and Some Welfare Effects." *Energy Economics, 31(2)*, 269-277. doi:10.1016/10.1007/s1 1149-006-9008-6.

Sioshansi, R. & Oren, S. S. (2007). "How Good are Supply Function Equilibrium Models: An Empirical Analysis of the ERCOT Balancing Market." *Journal of Regulatory Economics, 31(1)*, 1-35. doi: 10.1007/s1

1149-006-9008-6.

Stoddard, L., Abiecunas, J. & O'Connell, R. (2006). *Economic, Energy, and Environmental Benefits of Concentrating Solar Power in California.* NREL/SR-550-39291. Golden, CO: National Renewable Energy Laboratory.

US DOE (May 2008). *20% Wind Energy by 2030: Increasing Wind Energy's Contribution to U.S. Electricity Supply.* DOE/GO-102008-2567. Washington, DC: U.S. Department of Energy.

End Notes

[1] The HTF in a CSP plant will also have some thermal inertia that can help the CSP plant "ride out" a brief reduction in sunlight from a passing cloud.

[2] This is described in detail in section 4.3.

[3] A CSP plant can be designed with a fossil-fueled backup system. With such a design, natural gas or another fuel can be used to supplement solar thermal energy. Because our interest is in renewable resources, we focus on a pure CSP plant.

[4] See Gilman et al. (2008) for a detailed description of SAM and its capabilities.

[5] For purposes of comparison, the default heuristic dispatch rule in SAM is able to capture between 87% and 94% of the operating profits that our MIP model earns with a CSP plant with a solar multiple of 2.0 and six hours of TES.

[6] Available at http://rpm.nrel.gov/

[7] Solar direct normal irradiance in 2005 at the sites we consider was approximately 1-4% below the average over the period 1998-2006.

[8] Based on tests conducted of storage efficiencies from the Solar Two CSP plant in California

[9] 58.3 MWh-t corresponds to 20% of the energy required to run the power block at its design point of 110 MWh-e.

[10] Parasitics such as HTF pumps are also associated with operating the solar field. These parasitics are already accounted for and are netted out of the hourly thermal energy data we input from SAM into our dispatch model.

[11] The average given is the simple average, not a load-weighted average.

[12] Even in the ERCOT market, which has a much higher price cap than the CAISO market, Sioshansi and Oren (2007) suggest that the threat of regulatory intervention has served to suppress energy prices below levels that profit- maximizing behavior would support.

INDEX

A

B

C

D

E

F

G

H

I

J

K

L

M

N

O

P

Q

R

S

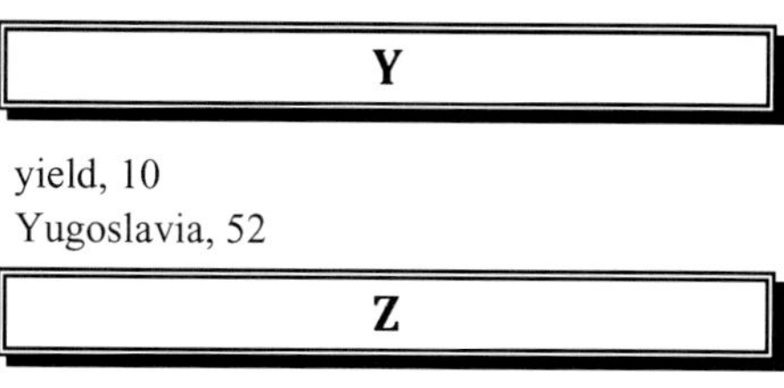

Y

Z